Thinking Spatially Using GIS

ISBN-13: 9781589481800
Suggested
grade level: 3–6

Thinking Spatially Using GIS presents engaging lessons that use fundamental spatial concepts and introduce GIS software to young students. Students are guided to find relative and absolute locations of map features, create maps, locate human and physical features on maps, discover and analyze geographic distribution patterns, and investigate changes over time. *Thinking Spatially Using GIS* provides a platform to tie in conventional 3–6 grade level topics, including lessons learned on world and U.S. geography, world exploration, and demographics.

Mapping Our World Using GIS

ISBN-13: 9781589481817
Suggested
grade level: 6 and up

Mapping Our World Using GIS invites middle and high school students to investigate GIS through lessons that require critical thinking and problem-solving skills. These lessons align with national teaching standards for geography, science, and technology in introducing and building skills in geographic inquiry, spatial thinking, and GIS. Topics covered include studying landforms, physical processes, ecosystems, climate, vegetation, population patterns and processes, human and political geography, and human and environment interaction.

Analyzing Our World Using GIS

ISBN-13: 9781589481824
Suggested
grade level: 9 and up

In *Analyzing Our World Using GIS,* students gain proficiency working with GIS and exploring geographic data. High school and college students will complete sophisticated workflows such as downloading and editing data and analyzing patterns on maps. Topics covered include analyzing economics by exploring education funding, demographics, and trade alliances; analyzing land and ocean effects on climate; and plate tectonics analysis to describe earthquake and volcanic activity.

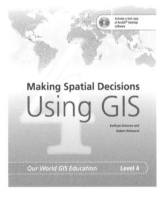

Making Spatial Decisions Using GIS

ISBN-13: 9781589481831
Suggested
grade level: 13 and up

Making Spatial Decisions Using GIS is intended for advanced high school, college, university, and technical school students. The book presents a wide variety of real-world settings for GIS analysis and decision making. Students learn methods for planning and executing GIS projects for more involved group investigations and independent study. Topics covered include analyzing hazardous emergency situations, using demographic data for analysis of population growth and urbanization trends, and using GIS to create reliable location intelligence that results in sound decisions.

Mapping Our World
Using GIS

Anita M. Palmer

Roger Palmer

Lyn Malone

Christine L. Voigt

ESRI PRESS
REDLANDS, CALIFORNIA

To our families, especially our husbands, Pat and Brian—
Thank you for your support, guidance, patience and love.

Lyn Malone, Christine L. Voight

To Joe Ferguson and Ann Judge, devoted teachers and eager students who
lost their lives on September 11, 2001—
Your passion and dedicated to geographic education lives on.

Ask for ESRI Press titles at your local bookstore or order by calling 1-800-447-9778. You can also shop online at www.esri.com/esripress. Outside the United States, contact your local ESRI distributor.

ESRI Press titles are distributed to the trade by the following:

In North America:
Ingram Publisher Services
Toll-free telephone: (800) 648-3104
Toll-free fax: (800) 838-1149
E-mail: customerservice@ingrampublisherservices.com

In the United Kingdom, Europe, and the Middle East:
Transatlantic Publishers Group Ltd.
Telephone: 44 20 7373 2515
Fax: 44 20 7244 1018
E-mail: richard@tpgltd.co.uk

Cover design and production   Jennifer Campbell
Interior design   Jennifer Campbell
Interior production   Savitri Brant
Image editing   Jay Loteria
Editing   Arthur Gelmis
Copyediting and proofreading   Tiffany Wilkerson
Permissions   Kathleen Morgan
Printing coordination   Cliff Crabbe and Lilia Arias

On the cover
World map by Michael Law. Cover data from ESRI Data & Maps 2006, courtesy of ArcWorld and National Geophysical Data Center.

Contents

Preface

We have written this book to provide you and your students with the opportunity to learn and use geographic information systems (GIS) to study geography and earth science concepts. The lessons are designed to enhance and extend your existing textbook content into the world of high-speed computing, vast databases, and the "supermaps" of geographic information systems. From these lessons, students will gain hands-on experience in using GIS technology as a powerful critical thinking tool that applies to a broad range of disciplines.

Our World GIS Education, Level 2: Mapping Our World Using GIS puts actual spatial data into the hands of your students—or more precisely, your computer—to analyze, interpret, and study various geography and earth science concepts such as tectonics, climate, population, political boundaries, economics, and human environment interaction. Displaying this data visually will allow students to make authentic, important decisions relating to real-life situations.

The lessons in this book will help teachers lead students through exercises in GIS mapmaking and will also illustrate and build skills in critical thinking and decision making. We have chosen topics that students will encounter in their geography and earth science classes; however, students will find these relevant and thought-provoking exercises will apply to many of their other courses as well. We know that students will enjoy using GIS to study geography and earth science content and will understand the importance of this powerful technology in solving everyday problems.

Mapping Our World Using GIS comes with a CD and a DVD. The CD contains the GIS data, student worksheets, and other documents you will need to complete the lessons. The DVD contains a 365-day trial version of ArcView 9.2 software. (Please see the installation instructions at the back of the book for system requirements.) This book provides information about online resources, and has a companion Web site, www.esri.com/ourworldgiseducation.

Our World GIS Education is a four-book series of comprehensive GIS instruction for students of all ages, from elementary school level to college undergraduates. The series builds on the solid foundation of *Mapping Our World: GIS Lessons for Educators*, the popular ESRI Press book geared to middle- or high-school classrooms. The exercises have been updated to the most current information available and the data has been updated to include 2007 and projected population statistics.

Mapping Our World Using GIS is the second volume in the *Our World GIS Education* series. It presumes no previous experience or skill in using geographic information systems and is appropriate for anyone from the middle school level and beyond. The introduction explains how to use this book and includes a list of steps for getting started. We have provided teachers notes, worksheets, and answer keys for your convenience. They will help you follow the exercises by logging answers to questions along the way and keeping track of the work to be completed.

These exercises have been used by thousands of teachers and students and we know your students will enjoy them as much as their peers have before them. With the enhanced data and exciting exercises included in this text, we are sure that both teachers and students alike will enjoy a new window into the world!

Anita Palmer, Roger Palmer, Lyn Malone, and Christine Voigt

About the authors

Anita M. Palmer is an education consultant based in Dallas, Texas. Anita and her husband Roger founded GIS Educational Technology Consultants (GISetc) in 1999 where they provide professional development for educators in the use of geospatial technologies in the K-16 classroom in the United States and abroad. She is a former high school teacher of technology in Carson City, Nevada. She designed and developed the GIS program and created the lauded advanced technology GIS course and internship program for underserved Hispanic young women. Anita co-authored the award winning book, *Mapping Our World: GIS Lessons for Educators* and the book *Community Geography: GIS in Action Teacher's Guide*. She also co-authored a lesson series about Louisiana Wetlands GIS for ArcLessons (http://www.esri.com/arclessons).

Roger Palmer is an experienced high school teacher of chemistry, physics, and environmental science and is an education consultant with GISetc. Fifteen years of teaching convinced of Roger of students' multiple learning styles, leading him to incorporate into his classes a mix of hands on laboratory, group projects, problem solving, reading sources, student competition, and field-work. The search for real world experiences led to monitoring local environments using electronic probeware and the need to represent data in a GIS. Student projects have been represented in local papers, national magazines, television news reports and at state and national conferences. As a GISetc consultant, Roger and his wife Anita have offered workshops in this approach to teachers across the United States and internationally.

Lyn Malone, a former teacher and cartographer, is an educational consultant specializing in the classroom application of geospatial technologies. In addition to *Mapping Our World*, Lyn is the co-author of *Community Geography: GIS in Action Teacher's Guide*. Lyn was Rhode Island's Social Studies Teacher of the Year in 1995 and has received awards from the National Council for Geographic Education, the National Geographic Society Education Foundation, American Forests, and ESRI for her innovative curriculum development and implementation In addition to writing, Lyn travels widely to conduct professional development workshops and institutes.

Christine L. Voigt is an award winning educator and free-lance curriculum writer from Dallas, Texas. Christine has a Bachelor of Arts in English from Texas Tech University and a Master of Science degree in Computer Education and Cognitive Systems from the University of North Texas. In addition to those degrees, she is a Teacher Consultant for National Geographic Society and a member of the ESRI K12 Authorized Teaching Program. Christine has co-authored two books with Anita Palmer and Lyn Malone, *Mapping Our World: GIS Lessons for Educators* and *Community Geography: GIS in Action Teacher's Guide*.

Acknowledgments

We thank the ESRI K–12 team of George Dailey, Charlie Fitzpatrick, and Angela Lee, whose vision, leadership, and hard work made this book a reality; Laura Bowden, K-16 program coordinator, who managed the project; Joseph Kerski, education industry curriculum development manager, who reviewed the book for geographical accuracy; Jennifer Campbell, who designed the cover and interior; Savitri Brant, who laid out the content; Arthur Gelmis, who edited the content; Tiffany Wilkerson, who copyedited the content; Jay Loteria and Scott McNair, who put together the digital media; Cliff Crabbe and Lilia Arias, who oversaw printing and production; Kathleen Morgan, Peter Schreiber, Lisa Horn, and Gail Hancock, who handled myriad legal issues; Mike Phoenix and Ann Johnson of the ESRI Higher Education team; Kelley Heider, who helped with a variety of administrative details; Judy Hawkins, ESRI Press acquisitions editor; Peter Adams, ESRI Press manager; Nick Frunzi, ESRI Educational Services director; and Jack Dangermond, ESRI founder and president, who made publication of this book a priority for his company.

We thank all those who contributed to *Mapping Our World: GIS Lessons for Educators*, the book from which this one is derived, including Kim Zanelli-English, Brian Parr, Tim Ormsby, Claudia Naber, and Eileen Napoleon. We thank the classroom teachers who reviewed and tested the lessons: Brad Baker of Bishop Dunne Catholic School, Dallas, Texas; Gerry Bell of Port Colborne High School, Port Colborne, Ontario; Shanna Hurt of Arapahoe High School, Littleton, Colorado; Marsha MacLean of the Redlands Unified School District, Redlands, California; Bart Manson and Joe Myszkowski of Red River High School, Grand Forks, North Dakota; Cathy Pleau of V. J. Gallagher Middle School, Smithfield, Rhode Island; Cynthia J. Ryan of Barrington Middle School, Barrington, Rhode Island; Herb Thompson of Greenspun Junior High School, Henderson, Nevada; and Patricia Walls of Taylor County Middle School, Grafton, West Virginia. We also thank Jim Trelstad-Porter, director of International Student Advising at Augsburg College, who interpreted Spanish data and correspondence with the Instituto Nacional de Estadística, Geografía e Informática in Mexico.

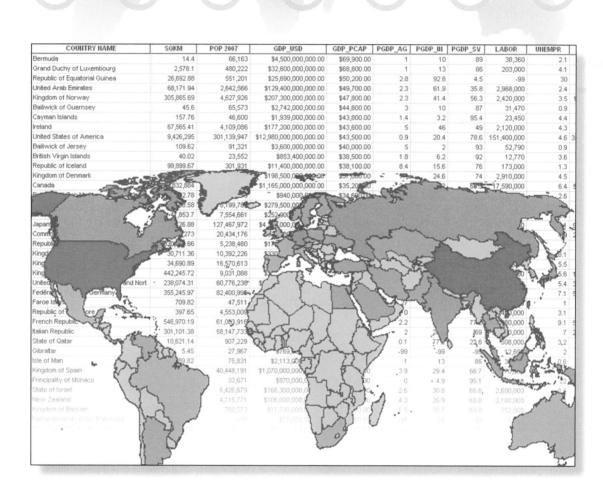

COUNTRY NAME	SQKM	POP 2007	GDP_USD	GDP_PCAP	PGDP_AG	PGDP_III	PGDP_SV	LABOR	UNEMPR
Bermuda	14.4	66,163	$4,500,000,000.00	$69,900.00	1	10	89	38,360	2.1
Grand Duchy of Luxembourg	2,578.1	480,222	$32,600,000,000.00	$68,800.00	1	13	86	203,000	4.1
Republic of Equatorial Guinea	26,692.88	551,201	$25,690,000,000.00	$50,200.00	2.8	92.6	4.5	-99	30
United Arab Emirates	68,171.94	2,642,566	$129,400,000,000.00	$49,700.00	2.3	61.9	35.8	2,968,000	2.4
Kingdom of Norway	305,865.69	4,627,926	$207,300,000,000.00	$47,800.00	2.3	41.4	56.3	2,420,000	3.5
Bailiwick of Guernsey	45.6	65,573	$2,742,000,000.00	$44,600.00	3	10	87	31,470	0.9
Cayman Islands	157.76	46,600	$1,939,000,000.00	$43,800.00	1.4	3.2	95.4	23,450	4.4
Ireland	67,565.41	4,109,086	$177,200,000,000.00	$43,600.00	5	46	49	2,120,000	4.3
United States of America	9,426,295	301,139,947	$12,980,000,000,000.00	$43,500.00	0.9	20.4	78.6	151,400,000	4.6
Bailiwick of Jersey	109.62	91,321	$3,600,000,000.00	$40,000.00	5	2	93	52,790	0.9
British Virgin Islands	40.02	23,552	$853,400,000.00	$38,500.00	1.8	6.2	92	12,770	3.6
Republic of Iceland	99,899.67	301,931	$11,400,000,000.00	$38,100.00	8.4	15.6	76	173,000	1.3
Kingdom of Denmark			$198,500,000,000.00	$37,000.00		24.6	74	2,910,000	4.5
Canada	832,884		$1,165,000,000,000.00	$35,200.00				17,590,000	6.4
	2.78		$940,000,000.00	$34,600.00					2.6
	58	8,199,783	$279,500,000.00						
	853.7	7,554,661	$252,000.00						
Japan	26.88	127,467,972	$4,000,000,000.00						
Comm	273	20,434,176							
Republic	66	5,238,460	$1						
Kingd	30,711.36	10,392,226	$330						8.1
King	34,690.89	16,570,613							5.5
King	442,245.72	9,031,088							5.6
United Kingdom and Nort	238,074.31	60,776,238							5.4
Federal Germany	355,245.97	82,400,996							7.1
Faroe Islands	709.82	47,511							1
Republic of	397.65	4,553,009			0			480,000	3.1
French Republic	546,970.19	61,083,916			2.2		77	80,000	9.1
Italian Republic	301,101.38	58,147,733			2		69	0,000	7
State of Qatar	10,621.14	907,229			0.1	77	22.6	508,000	3.2
Gibraltar	5.45	27,967	$769		-99	-99	-99	12,69	2
Isle of Man	89.82	75,831	$2,113,000		1	13	86		0.6
Kingdom of Spain		40,448,191	$1,070,000,000		3.9	29.4	66.7		
Principality of Monaco		32,671	$870,000,0		0	4.9	95.1		
State of Israel		6,426,679	$166,300,000,0		2.6	30.8	66.6	2,600,000	
New Zealand		4,115,771	$106,000,000		4.3	26.9	68.8	2,180,000	
Kingdom of Bahrain		708,573	$17,700,000		5	38.7	59.8	352,000	
Falkland Islands (Islas Malvinas)		2	$75,000,000						

Introduction

About the lessons

Mapping Our World Using GIS is a book of computer activities, data, and resources that can serve as a valuable supplement to topics you are already teaching in world geography, social studies, or earth science. Students will investigate global patterns of human and physical features, explore issues of concern to millions of people, analyze data from diverse regions, and develop skills essential for understanding a world characterized by vast quantities of raw information. Best of all, it will be fun!

The book comes with a one-year trial license of ArcGIS software. This means that your students will be using the same tools as the ones being used by professional planners, emergency response personnel, government agencies around the world, and businesses of all kinds.

Where to begin

We recommend you do the following before using this book in class:

1. Finish reading this section and skim through the book and the student workbook to locate the lessons, answer sheets, answer keys, worksheets, handouts, assessments, and rubrics.
2. Install and register the software on your computer and the student computers. (Refer to "Setting up the software and data" on page 5, and the detailed software installation guide at the back of the book.)
3. Install the teacher resources and student data on your computer from the Data and Resources CD. Install the data on student computers. (Refer to the detailed data installation guide at the back of the book.)
4. Work through module 1 by yourself.
5. Work through module 1 with your students. The two lessons in this module introduce students to the concept of GIS, basic ArcGIS software skills, and the steps of the geographic inquiry process.

Modules

After module 1, you and your students are free to explore modules 2 to 7 in any order you wish. You can teach each module or lesson independently of the others, and you can tailor the material to suit the specific needs of your class and curriculum. Each module consists of two lessons: a global perspective and a regional investigation.

How the materials are organized

This book contains teacher materials for each lesson, including the following:

- A lesson overview with a summary, materials list, time it will take to complete the lesson, learning objectives, a list of the main GIS tools and functions encountered in the activity, and key curriculum standards covered.
- Notes about teaching the lesson, including a "no-tech" activity to introduce lesson, tips on conducting the GIS activity, and how to conclude the lesson and give the optional assessment. Ideas for extending the lesson are listed as well.
- A list of the lesson components to be printed or copied for students (or used from the student workbooks).
- Assessment rubrics (see "Rubric-based assessments" on page 3).
- Answer keys for the student answer sheets and any supplemental worksheets.

The accompanying student workbook contains the activity sheets, answer sheets, and supplements that students will need to do the lessons. (It does not include the optional assessments.) You have several options for providing these materials to your students: you can order copies of the student workbook for your class; you can photocopy the student workbook pages that come with the book; or you can print them out from the files on the CD.

The Data and Resources CD has two main folders that can be installed separately: a teacher resources folder and a student data folder.

The teacher resources folder (OurWorld2_teacher) includes the following:

- Digital documents (PDF format) of the student activity sheets and supplements required by the lessons.
- Digital documents (Microsoft Word format) of the student answer sheets. These are provided in Word format so that you can add your own questions if desired. Optionally, you can ask students to type their answers. (This option does require screen space for the GIS windows and students will have to navigate back and forth between several programs at once.)
- Digital documents (PDF format) of some of the teacher materials in this book, such as assessment rubrics and answer keys.

The student data folder (OurWorld2) includes the lesson data that needs to be installed on student computers or in a network location where students can access (and modify) the data. The data is organized by module.

Curriculum standards

Each lesson teaches skills corresponding to National Geography Standards (*Geography for Life: The National Geography Standards 1994*). The applicable standards for middle school (grades 5–8) and high school (grades 9–12) are listed in each lesson overview. Matrices matching all lessons in the book to the National Geography Standards and to the National Science Education and National Technology Standards are also provided (see table of contents).

Reference handouts

The handouts "ArcMap Toolbar Reference," "ArcMap Zoom and Pan Tools," "Making Quality Maps," and "GIS Terms" can be used with any of the lessons. Consider giving these handouts to your students with module 1 and having the students save them for use with later lessons. The handouts can be printed from the Data and Resources CD. (They are also located at the back of the student workbooks, if you are using those.)

Rubric-based assessments

The lessons encourage students to explore a variety of geographic concepts and topics. A single letter or number grade won't be an accurate representation of the depth or completeness of their understanding of all concepts they've dealt with. The rubrics included with each lesson will allow you to evaluate student performance in a number of different ways. A learner may show mastery of one particular concept but perform another task at the introductory level. The rubrics will also help you provide specific feedback to your students, showing them exactly where they need additional assistance or practice. The four levels are defined as follows:

Exemplary: The student has gone above and beyond the standard. The student has a strong understanding of the concept and has the ability to mentor other students.

Mastery: This is the target level for all students. The student has a good understanding of the concept.

Introductory: The student has limited understanding of the standard or shows little evidence of understanding.

Does not meet requirements: The student does not show basic understanding of the standard.

The rubrics may be used as follows:

- Distribute a copy of the rubric to students when you return their evaluated work. Circle or highlight the student's level of achievement for each standard. This provides the greatest amount of feedback for the student on each standard. Use the back of the page to make additional comments.
- Use the rubric as a form of student self-evaluation. Give students an unmarked copy of the rubric and ask them to evaluate their own work.

The companion Web site

The book's Web site, www.esri.com/ourworldgiseducation, places a variety of GIS resources and other helpful information at your fingertips. For example, you'll want to check the Web site's "Resources by Module" section for specific resources, Web links, or changes when you get ready to use a particular lesson. Solutions to common problems and any significant changes or corrections to the book will also be posted here.

Taking it further

After your students have completed the lessons you have selected, you can do the following:

- Challenge them with lessons from *Our World GIS Education: Analyzing Our World Using GIS*.
- Have them put together a profile of your community and post it on the ESRI Community Atlas Web site: www.esri.com/communityatlas. Your school may be able to earn software through this program.
- Find out who's doing what with GIS near you and contact them for ideas. The following resources can help:
  - ESRI GIS Education Community, http://edcommunity.esri.com
  - ESRI Education User Conference, www.esri.com/educ
  - GIS Day Web site, www.gisday.com
  - GIS.com Web site, www.gis.com
  - Invite a GIS specialist from your city government or other local organization to do a presentation on GIS for your class.
  - Make GIS a permanent part of your classroom. Be sure to check with your district or state technology coordinator before you purchase an ArcView license for your school or classroom. A districtwide or statewide software license may already cover your school.

About the software and data

ArcGIS Desktop software

The lessons in this book use an ArcView license level of ArcGIS Desktop software. ArcGIS Desktop includes two applications: ArcMap and ArcCatalog. ArcMap is used to display and edit geographic data, perform GIS analysis, and create professional-quality maps, graphs, and reports. ArcCatalog is designed for browsing, managing, and documenting geographic data. Students interact only with ArcMap in this book's lessons, but you may find it useful to browse or preview the data using ArcCatalog as you are preparing to teach the lessons.

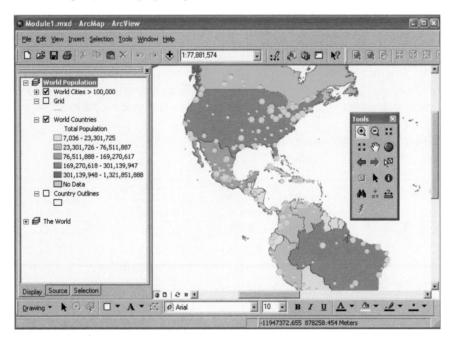

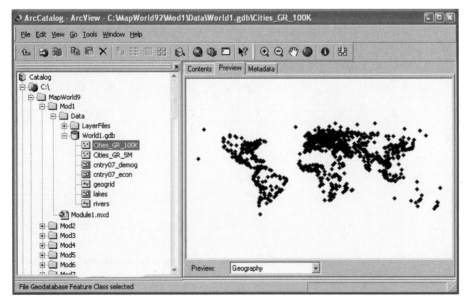

Setting up the software and data

This book comes with a DVD and a CD. The DVD contains a one-year trial edition of ArcView 9 for Microsoft Windows. The CD contains the data, student activity sheets, and other documents required for the lessons. Refer to the installation guides at the back of the book for detailed instructions on how to install the software and the data. The data license agreement is also found at the back of the book.

If you do not feel comfortable installing programs on your computer or your students' computers, please be sure to ask your school's technology specialist for assistance. The software and data on these disks need to be installed on your computer and all computers that the students will use to complete the GIS investigations. You may decide to install the student data on a school network drive. If you do choose this option, be aware that in some lessons (e.g., module 1, lesson 2) students must have their own copy of the data because the lesson requires them to modify it. The teacher resources should be installed on your computer, not on the student computers.

Once ArcView is installed there are a few things to do before the software is ready for students to use in the lessons:

Desktop shortcut. Consider creating an ArcMap shortcut icon on the desktop of each student's computer so that students can quickly locate and start the ArcMap program. If students will be using ArcCatalog, create an ArcCatalog shortcut icon as well.

Connection to the data. Most exercises instruct students to add data to their map documents at some point. To navigate to the exercise data, a connection to the drive or folder where the data is stored is necessary. You may want to make sure this folder connection is set up in advance on each student's computer, or you may direct students to create the folder connection themselves during the exercise. If you choose to create the folder connection yourself, "Installing the data and resources" at the back of the book explains how to do this using ArcCatalog. Otherwise, follow the instructions in module 1, lesson 2, step 4 to create the folder connection when students add data for the first time.

Student work folders. For some lessons you will need to set up student work folders where students can save their work. The activity instructions usually suggest that students include their initials in the name of the file they are saving, and advise them to ask their teacher where to save their work. See the "Teaching the lesson" section of individual lessons for more information.

File extensions. Instructions and graphics assume that students can see the file extensions, for example Global1.mxd or Cities.lyr. If your students cannot see file extensions (e.g., they see Global1 or Cities), you can either turn off this preference or have students ignore references to file extensions. Layer files, which sometimes have the same name as their data sources, are stored in separate folders to avoid confusion when file names are hidden. (To show file extensions, start ArcCatalog and go to Tools, then Options. Click the General tab and uncheck the Hide file extensions box.)

Moving or deleting data. The installation program installs the data for all modules. If for some reason you decide to remove part of the data, be sure to keep entire module folders intact. For example, if you only plan to teach lesson 1 of module 2, you should keep the entire module 2 folder. Lessons within a module often share some data, and map documents are set up to find data in specific locations within a module folder.

Metadata

Metadata (information about the data) describes the GIS data provided on the Data and Resources CD. The metadata includes a description of the data, where it came from, a definition for each attribute field, and other useful information. This metadata can be viewed in ArcMap or in ArcCatalog.

To view metadata using ArcMap, right-click a layer in the table of contents and select Data, then View Metadata. You may view information for each field in the layer by clicking the Attributes tab in the Metadata window and then clicking the field name. The default style sheet, FGDC ESRI, is satisfactory for most purposes, but if you want to view additional metadata detail, select ESRI Classic from the style sheet list.

Troubleshooting ArcGIS

Exercise instructions are written assuming the user interface and user preferences have the default settings. Unless students are working with a fresh installation of the software, however, chances are they will encounter some differences between the instructions and what they see on their screen. This is because ArcMap stores settings from a previous session. This could vary the size of the ArcMap window, the width of the table of contents, which toolbars are visible, where toolbars are located, or whether or not the map scale changes when the window is resized.

Normally such differences will not be a problem, but you should be prepared to help individual students if they question an instruction or want to know why their ArcMap looks different than their neighbor's. A list of commonly encountered troubles and their solutions can be found on this book's Web site, www.esri.com/ourworldgiseducation. You may want to print out this list for reference.

If you have questions related to installing the trial software or to report a problem or error with the lesson materials in this book, you can send e-mail to learngis@esri.com with your questions.

Geographic inquiry and GIS

Geography is the study of the world and all that is in it: its peoples, its places, its environments, and all the connections among them. Knowing where something is located, how its location influences its characteristics, and how its location influences relationships with other phenomena is the foundation of geographic thinking. Geographic inquiry asks you to see the world and all that is in it in spatial terms. Like other research methods, it also asks you to explore, analyze, and act upon the things you find.

The geographic inquiry process

Step	What to do
1. Ask a geographic question	Ask a question about spatial relationships in the world around you
2. Acquire geographic resources	Identify data and information that you need to answer your question
3. Explore geographic data	Turn the data into maps, tables, and graphs, and look for patterns and relationships
4. Analyze geographic information	Determine what the patterns and relationships mean with respect to your question
5. Act on geographic knowledge	Use the results of your work to educate, make a decision, or solve a problem

The five steps of geographic inquiry are addressed in detail in module 1, lesson 2. In the other lessons, these steps are implicit, and you will naturally integrate geographic inquiry into the process of doing the exercises throughout the book.

What is GIS?

Chances are that GIS technology has already touched your life. If you flipped on a light switch today, chances are that GIS was used to help make sure the electricity was there to light up the room. When you drove down a highway, chances are that GIS was used to keep track of the signs and streets along the way. If you received a delivery, chances are that GIS helped the driver find the way to your house. If you bought fresh vegetables, chances are that GIS helped manage the land and calculate the fertilizer needed for the crop. If you looked at a map on the Internet, chances are that GIS had a hand in that too.

A geographic information system (GIS) uses computers and software to organize, develop, and communicate geographic knowledge. In simple terms, GIS takes the numbers and words from the rows and columns in databases and spreadsheets and puts them on a map.

FIPS_CNTRY	CNTRY_NAME	POP2007
UV	Burkina Faso	14,326,203
GV	Guinea	9,947,814
PU	Guinea-Bissau	1,472,041
MI	Mali	11,995,402
		3,270,065
		12,521,851
		6,144,562
		1,688,359
		496,374
		4,906,585
		76,511,887
		2,874,127
		42,292,929
		30,262,610

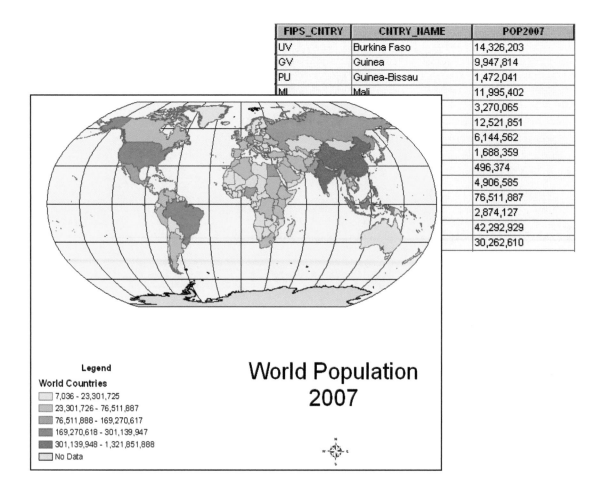

Legend
World Countries
- 7,036 - 23,301,725
- 23,301,726 - 76,511,887
- 76,511,888 - 169,270,617
- 169,270,618 - 301,139,947
- 301,139,948 - 1,321,851,888
- No Data

World Population 2007

Visualizing information

The vast amounts of information available today require powerful tools like GIS to help people determine what it all means. GIS can make thematic maps (maps coded by value) to help illustrate patterns. To explore cities at risk for an earthquake, you might first make a map of where earthquakes have already occurred. You could then code the earthquakes by magnitude. You might use one color for those that were strong and a second color for those that were weak. By analyzing the patterns, you will be able to find an answer to your question about cities at risk. You will pursue this inquiry in module 2.

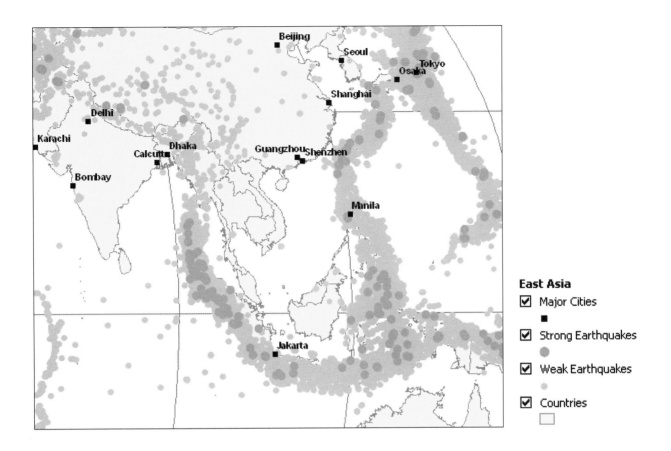

East Asia

☑ Major Cities
   ■

☑ Strong Earthquakes
   ●

☑ Weak Earthquakes
   ·

☑ Countries
   ▢

Putting it all together

GIS is a tool that can simplify and accelerate geographic investigations. Like any tool, GIS has no answers packed inside it. Instead, for those who use the tool and the process of geographic inquiry, it provides a means to discover pathways through our remarkable world of unending geographic questions.

ACQUIRE *geographic resources*

ASK *geographic questions*

EXPLORE *geographic data*

ACT on *geographic knowledge*

ANALYZE *geographic information*

Correlation to National Geography Standards

		Module 1		Module 2		Module 3		Module 4		Module 5		Module 6		Module 7	
	Standard	**L1**	**L2**	**L1**	**L2**	**L1**	**L2**	**L1**	**L2**	**L1**	**L2**	**L1**	**L2**	**L1**	**L2**
1	How to use maps and other geographic representations, tools, and technologies to acquire, process, and report information from a spatial perspective	●	●	●	●	●	●	●	●		●	●	●	●	●
2	How to use mental maps to organize information about people, places, and environments in a spatial context														
3	How to analyze the spatial organization of people, places, and environments on Earth's surface		●							●					
4	The physical and human characteristics of places	●	●	●	●		●				●			●	
5	That people create regions to interpret Earth's complexity					●									
6	How culture and experience influence people's perceptions of places and regions														
7	The physical processes that shape the patterns of Earth's surface			●	●	●									●
8	The characteristics and spatial distribution of ecosystems on Earth's surface														
9	The characteristics, distribution, and migration of human populations on Earth's surface									●					
10	The characteristics, distribution, and complexity of Earth's cultural mosaics														
11	The patterns and networks of economic interdependence on Earth's surface											●	●	●	
12	The processes, patterns, and functions of human settlement							●							
13	How the forces of cooperation and conflict among people influence the division and control of Earth's surface									●	●		●		
14	How human actions modify the physical environment														
15	How physical systems affect human systems			●	●	●									●
16	The changes that occur in the meaning, use, distribution, and importance of resources														
17	How to apply geography to interpret the past							●							
18	How to apply geography to interpret the present and plan for the future								●	●	●	●		●	●

Source: Geography Education Standards Projects. 1994. *Geography for Life: National Geography Standards 1994.* Washington, D.C.: National Geographic Research and Exploration.

Correlation to National Science Education Standards

	Standard	Module 1		Module 2		Module 3		Module 4		Module 5		Module 6		Module 7	
		L1	L2	L1	L2	L1	L2	L1	L2	L1	L2	L1	L2	L1	L2
	Unifying Concepts and Processes	All	All	All	All	All	All							All	All
A	Science as Inquiry		MH	MH	MH	MH	MH							MH	MH
B	Physical Science			MH	MH	M									M
C	Life Science						M							H	
D	Earth and Space Science			MH	MH	MH	MH							MH	
E	Science and Technology	MH	MH												
F	Science in Personal and Social Perspectives			MH	MH		MH	MH			MH			MH	MH
G	History and Nature of Science														

M = Middle school

H = High school

Standards reprinted with permission from National Science Education Standards. Copyright 1996 by the National Academy of Sciences. Courtesy of the National Academy Press, Washington, D.C.

Correlation to National Technology Standards

Standard	Module 1		Module 2		Module 3		Module 4		Module 5		Module 6		Module 7	
	L1	L2	L1	L2	L1	L2	L1	L2	L1	L2	L1	L2	L1	L2
1	●	●	●	●	●	●	●	●	●	●	●	●	●	●
2			●	●							●	●	●	●
3	●	●	●	●	●	●	●	●	●	●	●	●	●	●
4	●	●	●	●	●	●	●	●	●	●	●	●	●	●
5						●		●		●				
6	●	●	●	●	●	●	●	●	●	●	●	●	●	●

Sources: National Education Technology Standards for Students. 2007. International Society for Technology in Education.

MODULE 1

Geographic inquiry in ArcMap

Lesson 1: The basics of ArcMap

This lesson introduces the basic concepts and tools of GIS. It will guide students in navigating the computer to find ArcMap documents and data and acquiring fundamental GIS skills such as manipulating layers, zooming in and out, and identifying the attributes of geographic features.

Lesson 2: The geographic inquiry process

Students will learn the steps of geographic inquiry: ask, acquire, explore, analyze, and act. To test a geographic hypothesis, they will calculate people/phone line ratios and develop a plan of action based on their findings. They will begin to gain awareness of links between GIS and scientific inquiry, public policy, and economics.

Module 1 introduces basic skills on which the exercises in the rest of the book are based.

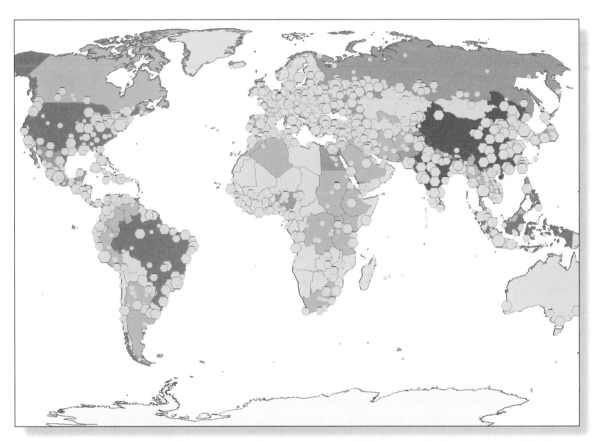

This GIS map has two layers: world countries classified by population and cities with populations greater than 100,000.

The basics of ArcMap

Lesson overview

This lesson introduces the basic concepts and tools of GIS. It will guide students in navigating the computer to find ArcMap documents and data and acquiring fundamental GIS skills such as manipulating layers, zooming in and out, and identifying the attributes of geographic features.

Estimated time

Two 45-minute class periods

Materials

The student worksheet files can be found on the Data and Resources CD. Install the teacher resources folder on your computer to access them.

Location: OurWorld_teacher\Module1\Lesson1
* Student PDF: M1L1_student.pdf
* Student answer sheet: M1L1_student_answer_sheet.doc

Objectives

After completing this lesson, a student is able to do the following:

* Understand the basic concept of a GIS
* Use a basic ArcMap skill set to build a map
* Print maps

GIS tools and functions

Identify a selected feature on the map

Zoom in to a desired section of the map or to the center of the map

Zoom to the full extent of the map

Clear selected features

Zoom out to a desired section of the map or to the center of the map

 Find a feature in a layer and identify it

 Move the map to bring a different portion of it into view

 Get help about a button

- Browse information about map features using MapTips
- Turn layers on and off
- Expand and collapse layers
- Use the mouse to zoom and pan
- Activate a data frame
- Change the order of the table of contents to change the map display
- Create a bookmark for a map extent and return to it later

National Geography Standards

Standard	Middle school	High school
1 How to use maps and other geographic representations, tools, and technologies to acquire, process, and report information from a spatial perspective	The student knows how to make and use maps, globes, charts, models, and databases to analyze spatial distributions and patterns	The student knows how to use technologies to represent and interpret Earth's physical and human systems
4 The physical and human characteristics of places	The student understands how technology can shape the characteristics of places	The student understands how the physical and human characteristics of place can change

Teaching the lesson

Introducing the lesson

Begin by helping students understand that each map in a GIS has database information tied to it. In other words, a GIS map takes the numbers and words from databases and ties them to a location. The map shows patterns in the data we might not otherwise see. This book's Data and Resources CD has a 2.5-minute video entitled "What is GIS?" that provides a basic description of GIS.

Student activity

We recommend that you complete the activity yourself before presenting the lesson in class. Doing so will allow you to modify the activity to accommodate the specific needs of your students. If they will not be working on individual computers, be sure to explain any necessary modifications.

Explain to students that in this activity they will begin to learn the basic ArcMap skills they will need to explore and navigate GIS maps. As they navigate through the lesson, they will be asked

questions that will help keep them focused on key concepts. Some questions will have specific answers, while others will require creative thought.

In questions 39, 44, 45, and 48, students are asked to use a calculator to divide two large numbers; some of these have 9 or 10 digits. If your students have handheld calculators that only allow 8-digit numbers, have them use the calculator accessory on their computer.

The following are things to look for while the students are working on this activity:

- Are students thinking spatially?
- Are students answering the questions?
- Are students using a variety of menus, buttons, and tools to answer the questions?
- Are students able to use the legends to interpret the data in the table of contents?
- Are students able to print out a map on the printer?

Concluding the lesson

Conduct a brief discussion on the tools the students used in their investigation of country populations and cell phones. Ask the students if they were surprised by their calculations of people per cell phone. Why or why not? Remind the students that there is a direct connection between the map and the attribute table. Have the students give an example of how that was illustrated in the lesson.

See the "Resources by Module" section of this book's Web site—www.esri.com/ourworldgiseducation—for print, media, and Internet resources that educate the public on the uses of GIS.

Answer key

Step 3: Work with layers

Q1.   Which layers are not visible on the map but are turned on in the table of contents?
Rivers, Lakes

Q2.   What happened on your map? The lakes show up on the map.

Q3.   What happened on your map? The rivers show up on the map.

Q4.   What would happen if you dragged Rivers under Land Areas? The rivers will disappear
again; or the rivers will be covered up by the Land Areas layer.

Step 4: Change the active data frame

Q5.   What is the name of the layer that is turned on in the World Population data frame? World
Countries

Step 6: Obtain information about a country

Q6.   What is the fourth listing under Field? CNTRY_NAME

Q7.   What is the fifth listing in this column? LONG_NAME

Q8.   What is the final listing in this column? MOBPHNS_YR

Q9.   What do you guess the field SQMI stands for? Square miles

Q10. What is the number to the right of the field SQMI? 2,970,758.75

Step 7: Compare Identify data with table data

Q11. Which row in the table has the attributes for Australia? The tenth row (OBJECTID 240)

Q12. Where are these field names displayed in the table? Across the top of the table; or as column
headings in the table

Q13. How many square miles of land does Australia cover? 2,970,758.75

Q14. Give a brief explanation of the relationship between the Identify window and the table. The
Identify window has the same items of information that are in the table. The Identify
window has information about only one country, while the table has all the countries.
(Students may not pick up on the second part just yet.)

Step 8: Explore city data on the world map

Q15. Use the Identify tool to find the names and countries of any two cities.

Answers will vary. Possible answers include:

City	Country
Zaragoza	Spain
Hamburg	Germany

Step 9: Explore Europe with an attribute table

Q16. What is the name of the table you opened? Attributes of World Countries

Q17. What country is listed in the first row of the table? Falkland Is.

Q18. What country is listed in the last row of the table? Serbia

Q19. What happens to the map when you click on these rows in the table? The three additional countries are outlined in blue.

Q20. What happens to Poland and the other countries that were highlighted? The blue outlines disappear.

Q21. Did you see the United States become outlined in blue on the map? If not, why not? Most students will not be able to see the United States outlined in blue because it isn't in the part of the world they are looking at. Some students may see the United States if they didn't zoom in enough or if they have a large screen.

Q22. Why can you see the United States now but not in the previous step? The instructions had the students zoom in too close on Europe for them to be able to see the United States. Once they zoomed to the full extent of the map, they were able to see the United States.

Step 10: Practice identifying features

Q23. What do you see on your map? South America only

Q24. What country is it? Brazil

Q25. What was this country's total population in 2007? 190,010,647

Q26. What city is it? Manaus

Q27. What population class is this city in? 1,000,000 to 5,000,000

Q28. What are the names of these two large cities? São Paulo, Rio de Janeiro

Q29. What population class are these cities in? 5,000,000 and greater

Step 11: Practice zooming out

Q30. What does your map look like? It's small or reduced in size.

Q31. Which button could you use to return your map to full size? Full Extent (the button that looks like a globe)

Step 12: Practice finding a feature

Q32. How many square kilometers in area is Sudan? 2,496,340

Q33. How many people lived in Sudan in 2007? 42,292,929

Q34. Compute the number of people per square kilometer. 16.94 people per square kilometer

Q35. Does the number of people per square kilometer seem low or high? It is a low number. Students may not be able to answer this question without looking at other countries. The United States has approximately twice the number of people per square kilometer, and some European countries have more. For instance, France has 7 times as many people per square kilometer, and Italy has 11 times the number of people per square kilometer. Students should wonder why this is such a low number. This is a good example of how geographic analysis can lead to further questions.

19

Step 13: Zoom to a feature and create a bookmark

Q36. Is Qatar a large country or a small one? A small one

Q37. How many people lived in Qatar in 2007? 907,229

Q38. How many cell phones did they have? 854,900

Q39. How many people were there for every cell phone in Qatar? 1.06 people per cell phone

Q40. What large country is directly west of Qatar? Saudi Arabia

Step 14: Explore the World Population map further

Q41. What boot-shaped country do you see on the map? Italy

Q42. What was the 2007 population of that country? 58,147,733

Q43. How many cell phones did that country have? 72,200,000

Q44. How many people were there for every cell phone in that country? 0.8 people per cell phone

Q45. How many cell phones were there per person in that country? 1.24 cell phones per person

Q46. What was the 2007 population of Japan? 127,467,972

Q47. How many cell phones did Japan have? 94,745,000

Q48. How many people were there for every cell phone in Japan? 1.35 people per cell phone

Q49. What happened to Qatar? It is no longer outlined in blue (the blue outline disappeared).

Step 16: Label and print a map

Q50. Where do you think these labels come from? The labels come from the country name attribute field (CNTRY_NAME) in the World Countries layer.

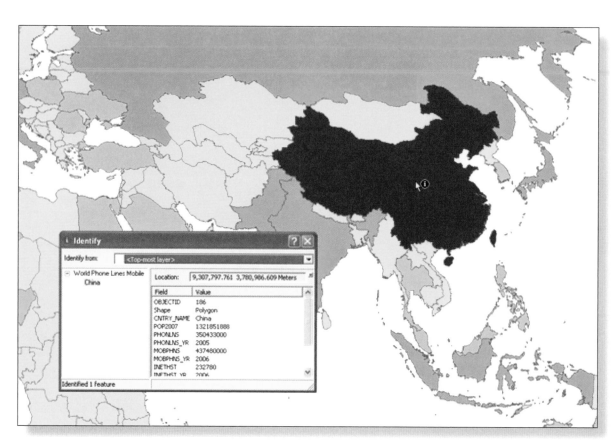

Data about telephone infrastructure in China is displayed using the Identify tool to access the "information behind the map."

The geographic inquiry process

Lesson overview

Students will learn the steps of geographic inquiry: ask, acquire, explore, analyze, and act. To test a geographic hypothesis, they will calculate people/phone line ratios and develop a plan of action based on their findings. They will begin to gain awareness of links between GIS and scientific inquiry, public policy, and economics.

Estimated time

Two 45-minute class periods

Materials

The student worksheet files can be found on the Data and Resources CD. Install the teacher resources folder on your computer to access them.

Location: OurWorld_teacher\Module1\Lesson2
- Student PDF: M1L2_student.pdf
- Student answer sheet: M1L2_student_answer_sheet.doc
- Student supplement: M1L2_supplement.pdf
- Student assessments: M1L2_assessment.pdf

Objectives

After completing this lesson, a student is able to do the following:

- Understand the basic concept of a GIS
- Use a basic ArcMap skill set to build a map
- Use the five-step geographic inquiry model
- Print maps

GIS tools and functions

 Add a layer to the map

 Connect to a folder containing GIS data

ⓘ Identify a feature on the map

◉ Zoom to the full extent of all the layers

▣ Clear selected features

🔍 Find a feature in a layer and identify it

- Turn layers on and off
- Expand and collapse layers
- Use the mouse to zoom and pan
- Calculate values for a field

National Geography Standards

Standard	Middle school	High school
1 How to use maps and other geographic representations, tools, and technologies to acquire, process, and report information from a spatial perspective	The student knows how to make and use maps, globes, charts, models, and databases to analyze spatial distributions and patterns	The student knows how to use technologies to represent and interpret Earth's physical and human systems
3 How to analyze the spatial organization of people, places, and environments on Earth's surface	The student understands that places and features are distributed spatially across Earth's surface	The student understands how spatial features influence human behavior
4 The physical and human characteristics of places	The student knows how different human groups alter places in distinctive ways	The student understands how the physical and human characteristics of place can change

Teaching the lesson

Introducing the lesson

This lesson is a continuation of lesson 1, so students should have completed lesson 1 before starting this one. The activity is structured on the geographic inquiry process. Read "Geographic inquiry and GIS" in the front section of the book to become familiar with this process.

Introduce the lesson with "The geographic inquiry process" supplement. Have a brief discussion with your students about thinking geographically.

Student activity

We recommend that you complete the activity yourself before presenting the lesson in class. Doing so will allow you to modify the activity to accommodate the specific needs of your students. If they will not be working on individual computers, be sure to explain any necessary modifications.

Explain to students that in this lesson they will practice the ArcMap skills they acquired in the first lesson and be introduced to more advanced skills while exploring the geographic inquiry process.

Students will be asked to calculate a field in an attribute table, which results in permanent changes to the data. Each student must therefore have their own copy of the module 1 folder.

The last two sections of the lesson do not involve the computer, and you may wish to assign them as homework.

The following are things to look for while the students are working on this activity:

- Are students thinking spatially?
- Are students answering the questions?
- Are students using a variety of menus, buttons, and tools to answer the questions?
- Are students able to use the legends to interpret the data in the table of contents?
- Are students able to print out a map on the printer?

Concluding the lesson

Conduct a brief discussion in which you ask students to brainstorm ideas about how GIS can be used in everyday life or how they could use GIS in their daily school assignments or classes. Ask them to describe the geographic inquiry process as well as share how comfortable they are with using ArcMap software. This discussion should also include an overview of which buttons they have used, any ArcMap operations that were confusing, and printer operations.

Middle school assessment. Students will have to create and print a map. They will be expected to turn on three layers and zoom in to a location of their choosing, use the Identify tool to obtain three pieces of data about that area, and write a brief paragraph explaining how they created their maps.

High school assessment. Students will have to create and print a map. They will be expected to turn on three layers and zoom in to a location of their choosing, use the Identify tool to obtain three pieces of data about that area, write a brief paragraph explaining how they created their maps, and describe what they learned geographically about the area on their map.

Extending the lesson

Challenge students to try the following:

- Create several maps with data from other folders on the data CD to answer a question. Make connections between the data to make sure the maps are meaningful.
- Calculate cell-phone density (the number of people per cell phone) using the World Phone Lines Mobile layer. Use maps and tables to explore how the use of mobile telephone technology may or may not have changed how a country develops its telephone land lines.
- Suggest other layers that might help explain the connections between telephone technology advances and sociopolitical issues like GDP, education, and health care.
- Develop a plan or outline for how to use GIS to fulfill a current class assignment.
- Choose a country in the news and use GIS skills to find the country and to study the data associated with it.

Answer key

Step 3: Ask a geographic question and develop a hypothesis

Q1. What makes this a geographic question? Answers will vary. Students should recognize that the question involves the distribution of something (phone lines) in different places (countries).

Q2. Write a hypothesis that answers the geographic question. The number of phone lines increases proportionately with the number of people in the world's most populous countries. Or, the number of phone lines does not increase proportionately with the number of people.

Step 4: Add a layer to your map

Q3. What other attribute of countries do you need in order to investigate your hypothesis? Number of phone lines

Q4. What is the name of the layer that has been added to your table of contents? World Phone Lines Land

Step 5: Explore the World Phone Lines Land map

Q5. What color in the legend for the World Phone Lines Land layer indicates countries with the fewest phone lines? Yellow

Q6. What color indicates countries with the most phone lines? Dark brown

Q7. What color indicates countries with no data available for this layer? Gray

Q8. What other layer in your map has a graduated color legend? World Countries

Q9. Which two countries have the most phone lines? China and the United States

Q10. On which continent are most of the countries with the fewest phone lines? Africa

Q11. Which two countries have the largest populations? China and India

Q12. Name the two countries that are in the same population class (color) as the United States. Brazil and Indonesia

Q13. Which of the two countries, if any, are in the same phone line class (color) as the United States? Neither of the countries

Q14. What two fields might help in answering the geographic question? POP2007, PHONLNS

Step 6: Investigate the relationship between phone lines and population in China

Record the answers to questions Q15–Q17 in the table below.

Q15. What was the population of China in 2007? See the table below.

Q16. How many phone lines did China have in 2007? See the table below.

Q17. What was the number of people per phone line in China? See the table below.

Country	Population	Phone lines	People/phone line
China	1,321,851,888	350,433,000	3.77

Step 8: Investigate the relationship between population and phone lines for all countries

Q18. What is the number of people per phone line (PHON_DEN) for China? 3.77

Q19. Does this number agree with the value you calculated in Q17? Yes

Q20. For each country in the table below, what is the population, number of phone lines, and number of people per phone line? The first country, China, is already filled in for you.

Country	Population	Phone lines	People/phone line
China	1,321,851,888	350,433,000	3.77
India	1,129,866,154	49,750,000	22.71
United States	301,139,947	268,000,000	1.12
Indonesia	234,693,997	12,772,000	18.38
Brazil	190,010,647	42,382,000	4.48
Pakistan	169,270,617	5,162,798	32.79
Russia	141,377,752	40,100,000	3.53
Japan	127,467,972	58,780,000	2.17

Step 9: Analyze the results of your research

Q21. In the table below, the column on the left ranks the countries by population from highest to lowest. In the column on the right, rank the countries from the lowest number of people per phone line to the highest number of people per phone line, using the data from Q20. Then draw lines connecting the name of each country in one column with the name of the same country in the other column.

Ranked by population (highest to lowest)	Ranked by people/phone line (lowest to highest)
China	United States
India	Japan
United States	Russia
Indonesia	China
Brazil	Brazil
Pakistan	Indonesia
Russia	India
Japan	Pakistan

Q22. Which country has the fewest people per phone line? United States
What is the number of people per phone line in this country? 1.12

Q23. How does the country in Q22 rank in population size among the eight countries in your table? It's the third largest in population.

Q24. Which country has the most people per phone line? Pakistan
What is the number of people per phone line in this country? 32.79

Q25. How does the country in Q24 rank in population size among the eight countries in your table? It's the sixth largest.

Q26. What is the population of Japan? 127,467,972
What is the number of people per phone line in Japan? 2.17

Q27. What country has the most phone lines? China
How does the number of people per phone line in this country compare with the numbers of people per phone line in the seven other countries in your table? It is fourth.

Q28. Russia and Pakistan have about the same number of people. Why do you suppose these two countries have such different numbers of people per phone line? What factors do you think contribute to this disparity? Possible answer: the countries have different cultures and different levels of economic development.

Q29. What is the answer to the geographic question? The number of phone lines does not vary proportionately with population.

Q30. How does your hypothesis (Q2) compare with your answer to the geographic question (Q29)? Answers will vary.

Step 10: Develop a plan of action

Q31. Use the information in the table in Q20 to describe the phone line situation in your chosen country.

Possible answers:

China has the highest number of phone lines of any country in the world, but it has almost four times as many people (over one billion) as phone lines. Brazil also has about four times as many people as phone lines; it ranks fifth among the eight countries, after the United States, Japan, Russia, and China.

Indonesia has 18.38 as many people as phone lines; it falls far behind other countries of similar size such as the United States (1.12 people per phone line) and Brazil (4.48 people per phone line).

The United States has 268 million phone lines—second-highest in the world; it leads the world in access to phone lines, with about as many phone lines as people.

Q32. Do you think that increasing the number of phone lines operating in your chosen country would improve the quality of life there? Why or why not? Answers will vary.

Q33. List three concerns you have about increasing the number of phone lines in your chosen country. Possible answers: cost, difficulty of providing access in rural areas or among the many islands of Indonesia, a preference for expanding more modern cell-phone infrastructure, or greater importance of other issues in the less developed countries (e.g., education, health care).

Q34. List two new geographic questions that you would like to investigate to help you develop a sound plan. Answers will vary but should include a geographic component.

Assessment rubrics

Middle school

Standard	Exemplary	Mastery	Introductory	Does not meet requirements
1 The student knows how to make and use maps, globes, graphs, charts, models, and databases to analyze spatial distributions and patterns	Creates and prints a focused GIS map with more than three themes	Creates and prints a focused GIS map with three themes	Creates and prints a focused GIS map with one or two themes	Has difficulty creating a map without assistance and does not print it out
3 The student understands that places and features are distributed spatially across Earth's surface	Identifies more than three pieces of information about the area of the world covered by his or her map and develops a geographic question based on that information	Identifies three pieces of information about a particular area of the world and develops a geographic question based on that information	Identifies one or two pieces of information about a particular area of the world and attempts to formulate a geographic question based on that information	Identifies some information about a place but does not create a geographic question based on the information gathered
4 The student knows how different human groups alter places in distinctive ways	Compares a chosen region to China or India using four or more characteristics such as technology, landscape, climate, population, and density	Compares a chosen region to China or India using two or three characteristics	Provides a vague comparison of a chosen region to China or India	Makes an incorrect comparison of a chosen region to China or India

This is a four-point rubric based on the National Standards for Geographic Education. The mastery level meets the target objective for grades 5–8.

High school

	Standard	Exemplary	Mastery	Introductory	Does not meet requirements
1	The student knows how to use technologies to represent and interpret the Earth's physical and human systems	Creates and prints a focused GIS map with more than three layers	Creates and prints a focused GIS map with three layers	Creates and prints a focused GIS map with one or two layers	Has difficulty creating the map without assistance and does not print it out
3	The student understands how spatial features influence human behavior	Identifies more than three pieces of information about the area of the world covered by his or her map and uses that information to develop a geographic question about the human impact on that place	Identifies three pieces of information about a particular area of the world and uses that information to develop a geographic question about the human impact on that place	Identifies one or two pieces of information about a particular area of the world and attempts to formulate a geographic question based on that information	Identifies some information about a place but does not create a geographic question based on the information gathered
4	The student understands how the physical and human characteristics of place can change	Describes three or four plausible connections between geographic characteristics and the character of a place; writes a clear and concise comparison that takes into account the physical and human characteristics of the two places	Describes one or two plausible connections between a place's geographic characteristics and its character; writes a clear and concise comparison of the two places	Briefly describes geographic characteristics of a place; identifies some common physical and human characteristics of the two places	Lists one or two characteristics of a chosen region but does not provide a comparison

This is a four-point rubric based on the National Standards for Geographic Education. The mastery level meets the target objective for grades 9–12.

MODULE 2

Geology

Lesson 1: The earth moves: A global perspective

Students will observe seismic and volcanic activity patterns around the world, analyze the relationships of those patterns to tectonic plate boundaries and physical features on the earth's surface, and identify cities at risk.

Lesson 2: Life on the edge: A regional investigation of East Asia

Students will investigate the East Asia portion of the Ring of Fire, where millions of people live with the daily threat of significant seismic or volcanic events. They will identify zones of tectonic plate subduction and populations at risk.

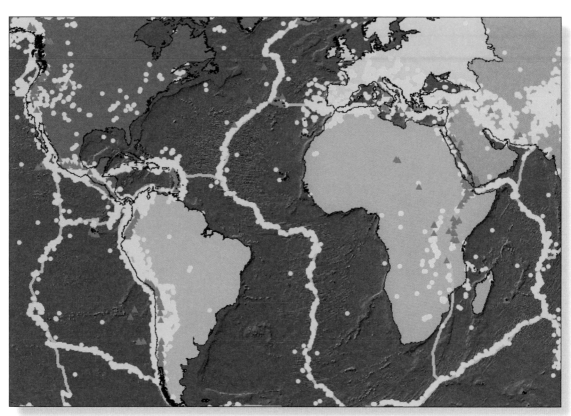

The global distribution pattern of earthquakes and volcanoes closely resembles tectonic plate boundaries.

The earth moves

A global perspective

Lesson overview

Students will observe seismic and volcanic activity patterns around the world, analyze the relationships of those patterns to tectonic plate boundaries and physical features on the earth's surface, and identify cities at risk.

Estimated time

Two 45-minute class periods

Materials

The student worksheet files can be found on the Data and Resources CD. Install the teacher resources folder on your computer to access them.

Location: OurWorld_teacher\Module2\Lesson1
- Student PDF: M2L1_student.pdf
- Student answer sheet: M2L1_student_answer_sheet.doc
- Student supplements: M2L1_supplement.pdf
- Student assessments: M2L1_assessment.pdf

Additional materials
- Large wall map of the world
- 100 adhesive dots or map pins in two colors
- Colored pins

Objectives

After completing the lesson, a student is able to do the following:

- Locate zones of significant seismic or volcanic activity
- Describe the relationship between zones of high seismic or volcanic activity and tectonic plate boundaries
- Identify cities threatened by earthquakes or volcanic eruptions

GIS tools and functions

ⓘ Identify a selected feature on the map

▣ Clear selected features

◕ Zoom to the full extent of the map

✚ Add layers to the map

⬏ Browse up one level

- Open the attribute table for a layer
- Sort data in ascending or descending order
- Select records in a table
- Select features based on an attribute
- Zoom and pan the map using the mouse

National Geography Standards

Standard	Middle school	High school
1 How to use maps and other geographic representations, tools, and technologies to acquire, process, and report information from a spatial perspective	The student knows and understands how to use maps to analyze spatial distributions and patterns	The student knows and understands how to use geographic representations and tools to analyze and explain geographic problems
4 The physical and human characteristics of places	The student knows and understands how to analyze the physical characteristics of places	The student knows and understands the changing physical characteristics of places
7 The physical processes that shape the patterns of Earth's surface	The student knows and understands how physical processes shape patterns in the physical environment	The student knows and understands the spatial variation in the consequences of physical processes across Earth's surface
15 How physical systems affect human systems	The student knows and understands how natural hazards affect human activities	The student knows and understands how humans perceive and react to natural disasters

Teaching the lesson

Introducing the lesson

Have students mark with the letters *V* and *E* eight locations on the Supplement map where they believe volcanoes are located and eight locations where they think earthquakes typically occur. Once students have done this individually, have them discuss their ideas in groups of three or four,

36

answer the questions in the Supplement as a group, and choose five volcano locations and five earthquake locations to mark on the large class map. (Use one color for earthquakes and the other for volcanoes.) After all the groups have marked their selections with adhesive dots or map pins, briefly discuss the patterns on the map.

Student activity

We recommend that you complete the activity yourself before presenting the lesson in class. Doing so will allow you to modify the activity to accommodate the specific needs of your students. If they will not be working on individual computers, be sure to explain any necessary modifications.

Students will use GIS to identify zones of volcanic and seismic activity and cities at risk.

The following are things to look for while the students are working on this activity:

- Are the students using a variety of tools?
- Are the students answering the questions?
- Are the students referring to their original maps and notes from group discussion?

Concluding the lesson

Engage students in a discussion of the observations and discoveries that they made during their exploration of the maps. Ask students to compare their initial ideas—as reflected by the pattern of colored dots on the class wall map—with the insights they acquired in the course of the activity. How have their ideas about earthquakes and volcanoes changed since the start of the lesson? Has this lesson raised any questions that they would like to explore further? How can GIS help world cities better prepare for seismic and volcanic events?

Middle school assessment. Students are asked to mark plate boundaries, physical features, and cities at risk on a paper map; to describe the relationships between plate boundaries and areas of frequent earthquake or volcanic activity; and reflect on their selections of cities at risk.

High school assessment. Students are asked to mark plate boundaries, physical features, and cities at risk on a paper map; to research a city's disaster preparedness plan; and to hypothesize on how selected physical features were created.

Extending the lesson

Challenge students to try the following:

- Create their own digital maps for the assessment, marking the required features and printing the map using the layout view
- To create a new layer in ArcMap that identifies natural hazards in your region
- Look at historical data for notable earthquakes and volcanic eruptions, incorporate this data as new layers, and predict the next significant eruption or seismic event
- Analyze fault line data included with this lesson (**World2.gdb\faults**) to see if it provides additional insight on locations and movement of plate boundaries
- Compare earthquakes in 2003 with earthquakes in 2007 and identify the locations of more recent earthquakes

See the "Resources by Module" section of this book's Web site—www.esri.com/ourworldgiseducation— for print, media, and Internet resources on the topics of earthquakes, volcanoes, and plate tectonics.

Answer key

Step 2: Analyze earthquake locations

Q1. Do earthquakes occur in the places you predicted? List the regions you predicted correctly for earthquake locations. **Answers will vary depending on student predictions.**

Q2. What patterns do you see on the map? **Answers will vary and can include that many earthquakes occur on the western coast of North and South America, along the eastern coast of Asia, and along the islands of the Pacific Rim. The pattern follows the Ring of Fire. They may also note a pattern of earthquakes down the center of the Atlantic Ocean and a string in the southern Atlantic, from South America eastward through to the Indian Ocean. Another string runs east and west through south-central Asia and southern Europe.**

Step 3: Sort and analyze earthquake magnitudes

Q3. How do the 20 selected locations compare to your map in supplement A? List three ways. **Answers will vary depending on student predictions. If students selected any spots around the Ring of Fire, then their predictions were close to reality.**

Step 4: Analyze volcano data

Q4. How do the volcano locations compare with your original predictions? List the regions of volcanic activity you predicted correctly. **Answers will vary depending on student predictions. If students selected any spots around the Ring of Fire, then their predictions were close to reality.**

Q5. What patterns do you see in the volcano locations, and how do they compare with the earthquake patterns? **The earthquake and volcano locations line up in similar patterns with the exception of the volcanoes in some areas of Africa.**

Step 5: Select all active volcanoes

Q6. What pattern do you see? **The majority of the volcanoes on the map are active, particularly around the Pacific Rim.**

Q7. Formulate a hypothesis as to why volcano eruptions and earthquakes happen where they do. **Answers will vary but should allude to the idea of plate tectonics and the fact that movement at plate boundaries causes disruptions on the earth's surface.**

Step 6: Identify active volcanoes on different continents

Q8. Use the Identify tool to find the names, elevations, and countries of three active volcanoes. **Possible answers: Banahao, 2,177 m, Philippines; Ibu, 1,340 m, Indonesia; Haku San, 2,702 m, Japan.**

M ● ● ○ ○ ○ ○ ○
L ● ○

Step 7: Add the plate boundaries layer

Q9.   Based on the locations of earthquakes and volcanoes, where do you think the plate boundaries are? Draw them on the Supplement map. Answers will vary.

Q10. Compare the actual plate boundaries to the ones you drew on the Supplement map. Record all similarities and differences. Answers will vary based on students' original hypotheses. If students drew the boundaries to follow the patterns of earthquakes and volcanoes, then they are on target.

Step 8: Add a layer file and an image

Q11. Are there any areas where physical features, plate boundaries, and seismic and volcanic activities overlap? Answers may include the eastern part of the map where the Philippine and Pacific plates meet, eastern Africa, the Mid-Atlantic Ridge, and the western coast of North and South America.

Q12. Write the names of physical features in the first column of the table below and label them on the Supplement map. Possible answers are listed in the table.

Physical feature	How the physical feature was created
Mid-Atlantic Ridge	Divergence of the South American and African plates and divergence of the North America and Eurasia plates
Aleutian Trench Aleutian Islands	Convergence of the North America and Pacific plates
Rocky Mountain Range Cascade Mountain Range Sierra Nevada Mountain Range Baja California	Convergence of the North America and Pacific plates
Sierra Madre Occidental Sierra Madre del Sur	Convergence of the North America, Pacific, and Cocos plates
Andes Mountain Range Peru–Chile Trench	Convergence of the Nazca and South America plates
Alps Mountain Range Atlas Mountains	Convergence of the Africa and Eurasia plates
East Africa Rift Valley	Divergence of the Africa, Arabia, and Somalia plates
Himalayas Tibetan Plateau	Convergence of the India and Eurasia plates
Mariana Trench	Convergence of the Philippine and Pacific plates

Q13. In the second column of the table above, write how you think each physical feature was created. Possible answers are listed in the table. Students can find additional physical features by consulting an atlas or a physical map of the world.

Step 9: Identify major cities at high or low risk for seismic or volcanic activity

Q14. List five high-risk cities and five low-risk cities. Possible answers are listed below.

High-risk cities	Low-risk cities
Tokyo, Japan	Minsk, Belarus
Reykjavik, Iceland	Lagos, Nigeria
San Francisco, United States	Winnipeg, Canada
Quito, Ecuador	Salvador, Brazil
Managua, Nicaragua	Madurai, India

M ● ● ○ ○ ○ ○ ○
L ● ○

Assessment rubrics

Middle school

Standard	Exemplary	Mastery	Introductory	Does not meet requirements
1 The student knows and understands how to use maps to analyze spatial distributions and patterns	Uses GIS to analyze volcanic and earthquake data to identify five cities most at risk for volcanic or seismic disasters, provides ample evidence for his or her choices, and draws conclusions on commonalities between the places	Uses GIS to analyze volcanic and earthquake data to identify five cities most susceptible to volcanic or earthquake disasters and provides ample evidence to support his or her decisions	Uses GIS to identify four or five cities most susceptible to volcanic or earthquake disasters and provides limited support for his or her decisions	Uses GIS to identify some cities susceptible to volcanic or earthquake disasters but does not provide evidence to support his or her choices
4 The student knows and understands how to analyze the physical characteristics of places	Identifies at least five physical features created by volcanic or seismic activity and all plate boundaries	Identifies three to five physical features created by volcanic or seismic activity	Identifies two or three physical features created by volcanic or seismic activity	Identifies one physical feature created by volcanic or seismic activity
7 The student knows and understands how physical processes shape patterns in the physical environment	Clearly describes the relationship between zones of high earthquake or volcanic activity and tectonic plate boundaries through the use of a variety of media	Describes the relationship between zones of high earthquake or volcanic activity and tectonic plate boundaries	Provides limited evidence of the relationship between zones of high earthquake or volcanic activity and tectonic plate boundaries	Does not show understanding of the relationship between zones of high earthquake or volcanic activity and tectonic plates
15 The student knows and understands how natural hazards affect human activities	Ranks five high-risk cities and provides ample evidence, including outside sources, for the rank order	Ranks five high-risk cities and provides a clear explanation for the rank order	Ranks five cities but does not provide an explanation for the rank order	Does not rank the cities

This is a four-point rubric based on the National Standards for Geographic Education. The mastery level meets the target objective for grades 5–8.

High school

Standard	Exemplary	Mastery	Introductory	Does not meet requirements
1 The student knows and understands how to use geographic representations and tools to analyze and explain geographic problems	Uses GIS to analyze volcanic and earthquake data to identify five or more cities at high risk for volcanic or earthquake activity and provides a detailed explanation of how he or she came to these conclusions	Uses GIS to analyze volcanic and earthquake data to identify five cities at high risk for volcanic or earthquake activity and provides a brief narrative stating the logic for their conclusions	Uses GIS to identify three to five cities at high risk for volcanic or earthquake activity but provides little explanation for the choices	Uses GIS to identify one to five high-risk cities but does not provide any explanation for the choices
4 The student understands the changing physical characteristics of places	Clearly describes through various media how plate tectonics shaped five physical features	Describes with ample evidence how plate tectonics shaped three to five physical features	Describes in some detail how plate tectonics shaped two or three physical features	Identifies a few surface formations but provides little description of how plate tectonics shaped them
7 The student knows and understands the spatial variation in the consequences of physical processes across Earth's surface	Clearly explains the correlation of the spatial distribution of volcanoes and earthquakes to plate boundaries and surface formations	Explains the correlation of the spatial distribution of volcanoes and earthquakes to plate boundaries	Identifies the correlation of volcano and earthquake locations to plate boundaries but has difficulty explaining the relationship	Identifies similarities in the spatial distributions of volcanoes and earthquakes but not the correlation to plate boundaries
15 The student knows and understands how humans perceive and react to natural disasters	Provides a research-based description of how three or more high-risk cities have prepared for a major volcanic or seismic event	Provides a research-based description of how two high-risk cities have prepared for a major volcanic or seismic event	Describes how a high-risk city has prepared for a major seismic or volcanic event and provides some evidence	Describes how a high-risk city has prepared for a major seismic or volcanic event but does not provide evidence

This is a four-point rubric based on the National Standards for Geographic Education. The mastery level meets the target objective for grades 9–12.

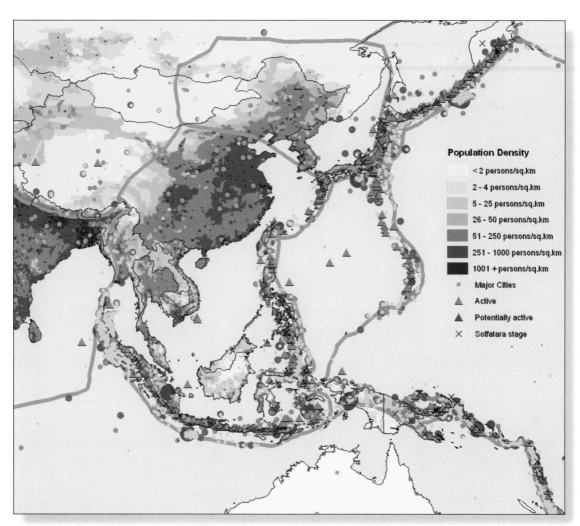

Population Density

< 2 persons/sq.km

2 - 4 persons/sq.km

5 - 25 persons/sq.km

26 - 50 persons/sq.km

51 - 250 persons/sq.km

251 - 1000 persons/sq.km

1001 + persons/sq.km

● Major Cities

▲ Active

▲ Potentially active

✕ Solfatara stage

Mapping population density and major cities with tectonic activity suggests where people are at risk.

Life on the edge

A regional investigation of East Asia

Lesson overview

Students will investigate the East Asia portion of the Ring of Fire, where millions of people live with the daily threat of significant seismic or volcanic events. They will identify zones of tectonic plate subduction and populations at risk.

Estimated time

Two to three 45-minute class periods

Materials

The student worksheet files can be found on the Data and Resources CD. Install the teacher resources folder on your computer to access them.

Location: OurWorld_teacher\Module2\Lesson2
- Student PDF: M2L2_student.pdf
- Student answer sheet: M2L2_student_answer_sheet.doc
- Student supplements: M2L2_supplement.pdf
- Student assessments: M2L2_assessment.pdf

Additional materials
- Colored pencils

Objectives

After completing the lesson, a student is able to do the following:

- Locate zones of significant earthquake and volcanic activities in East Asia
- Describe the relationship between zones of high earthquake or volcanic activity and tectonic plate boundaries
- Identify subduction zones along plate boundaries
- Identify densely populated areas that are most at risk for volcanic and/or seismic disaster

GIS tools and functions

Identify a feature to learn more about it

Zoom in on the map

Measure distances between points on the map

Add layers to the map

Pan the map to view different areas

- Label features on the map
- Turn layers on and off
- Change the order of the table of contents to change the map display

National Geography Standards

Standard	Middle school	High school
1 How to use maps and other geographic representations, tools, and technologies to acquire, process, and report information from a spatial perspective	The student knows and understands how to use maps and databases to analyze spatial distributions and patterns	The student knows and understands how to use geographic representations and tools to analyze and explain geographic problems
4 The physical and human characteristics of places	The student knows and understands how to analyze the physical characteristics of places	The student knows and understands the changing physical characteristics of places
7 The physical processes that shape the patterns of Earth's surface	The student knows and understands how to predict the consequences of physical processes on Earth's surface	The student knows and understands the spatial variation in the consequences of physical processes across Earth's surface
15 How physical systems affect human systems	The student knows and understands how natural hazards affect human activities	The student knows and understands how humans perceive and react to natural disasters

Teaching the lesson

Introducing the lesson

Provide a brief overview of the region of East Asia and the Ring of Fire. Emphasize East Asia's dense population. Ask students to draw on the Supplement map outlines of areas they believe to have the greatest risk for a major geophysical disaster. Allocate about 5 minutes for this task. Then, if time permits, ask some students to share their predictions with the class or divide students into small groups for discussion.

Student activity

We recommend that you complete the activity yourself before presenting the lesson in class. Doing so will allow you to modify the activity to accommodate the specific needs of your students. If they will not be working on individual computers, be sure to explain any necessary modifications.

Students will analyze population density, plate boundaries, volcanoes, and earthquake activity for the years 2004 through mid-2007, and they will identify areas of subduction along plate boundaries.

The following are things to look for when the students are working on this activity:

- Are the students answering the questions?
- Are the students referring to their original maps and notes from class discussion?
- Do some students complete the exercise quickly? If yes, refer to "Extending the lesson" for additional challenges.

Concluding the lesson

Lead a class discussion comparing student predictions of disaster-prone areas (Supplement) and their findings from the investigation. How closely did their predictions match what they learned?

Assessment. Students are asked to mark zones at risk for volcanic and seismic events on a paper map, to create a legend, and to analyze the map.

After students complete the assessment, you may want to solicit additional discussion:

- Have students identify what data was most important in determining risk zones in the assessment. One way to do this is to list types of data on the board and have the students rank them in order of importance.
- Discuss why some data was more helpful than other data.
- Ask students if there is other data that they think would be helpful and to explain why.

Extending the lesson

Challenge students to try the following:

- Create a buffer zone around a volcano near an East Asian city to determine the area of potential damage in case of a major eruption and then create a disaster plan for the city.
- Analyze historical data for the most notable earthquakes and volcanic eruptions, create new layers from this data, and make predictions on the next significant eruption or seismic event for East Asia.
- Research and report on the December 26, 2004, Sumatra-Andaman earthquake and its effects. Mark the earthquake location on the map using the Draw tools and create a map using the map document and data from this lesson to illustrate the report.
- Create a new layer showing locations of natural hazards in your local area.
- Create digital maps for the assessment and print them using the layout view.
- Present their maps to the class, comparing their predictions on Supplement maps to the results of their analysis on Assessment maps.

Check out the "Resources by Module" section of this book's Web site—www.esri.com/ourworldgis education—for print, media, and Internet resources on the topics of earthquakes, volcanoes, plate tectonics, and East Asia.

Module 2: Lesson 2

Answer key

Step 1: Open a map document and identify cities

Q1. Use the Identify tool to locate one city within each country listed in the table below and record that city's population. Possible answers are listed in the table.

City	Country	City population
Kunming	China	1,280,000
Delhi	India	7,200,000
Tokyo	Japan	23,620,000

Step 2: Look at population density and earthquake magnitudes

Q2. Use the Identify tool to locate two East Asian cities in areas where population density is greater than 250 people per square kilometer. Possible answers: Nanjing, China; Calcutta, India

Q3. Describe the general pattern of population density in East Asia. Areas of high population density are found in the southeastern part of the region. Specific areas students may mention include eastern China, the Indian subcontinent, southeast Asia peninsula, Japan, and some islands. The remainder of East Asia has generally low population density.

Q4. In general, where did earthquakes with a magnitude of ≥5 occur? Answers will vary but should mention that many of the earthquakes occurred along the islands of the Pacific Rim, with some occurring under the ocean.

Q5. Did these earthquakes occur near densely populated areas? Where? Yes: Japan, Indonesia, Philippines, Taiwan

Step 3: Measure the distance between active volcanoes and nearby cities

Q6. What is the closest distance you found between a volcano and a city? Record that city, the volcano, and the distance between them. Possible answer: City: Manado, Indonesia. Volcano: Mahawu. Distance: 7 kilometers

Q7. Are there many active volcanoes located close to cities? Yes

Q8. What patterns do you see in the locations of volcanoes, and how do they compare with the earthquake patterns? The patterns of earthquake and volcano locations are virtually the same, especially along the islands of the Pacific Rim (the western edge of the Ring of Fire).

Step 4: Look at plate boundaries

Q9. Record labels on the Supplement map. Refer to the Region2 map document for correct answers.

Step 5: Add an image file

Q10. On the Supplement map, draw the zones of subduction. Student maps should indicate subduction zones along the western boundary of the Pacific plate and the eastern and southern boundaries of the Indo-China plate.

Step 6: Investigate your map

Q11. Record the name of the plate you investigated. List three cities and three physical features in the vicinity. Possible answers are listed below.

Plate name: Amur Plate
Cities:

1.   Nagoya, Japan
2.   Pusan, South Korea
3.   Vladivostok, Russia

Physical features:

1.   Fuji-san (volcano)
2.   Haku san (volcano)
3.   Ocean trench at southern boundary

Assessment rubrics

Middle school

Standard	Exemplary	Mastery	Introductory	Does not meet requirements
1 The student knows and understands how to use maps and databases to analyze spatial distributions and patterns	Analyzes GIS data to create a paper map showing zones of high risk and low risk for volcanic and seismic activities	Analyzes GIS data to identify zones of significant seismic and volcanic activities	Uses GIS to identify the locations of major volcanoes and earthquakes	Has difficulty correctly identifying major volcanoes and earthquakes
4 The student knows and understands how to analyze the physical characteristics of places	Describes in detail the characteristics of Pacific Rim physical features and identifies the physical processes that created them	Clearly describes the characteristics of Pacific Rim physical features	Describes the characteristics of Pacific Rim physical features using little detail	Identifies Pacific Rim physical features but does not provide specific characteristics
7 The student knows and understands how physical processes shape patterns in the physical environment	Clearly describes the relationship between zones of high earthquake or volcanic activity and tectonic plate boundaries through a variety of media	Describes the relationship between zones of high earthquake or volcanic activity and tectonic plate boundaries	Provides limited evidence of the relationship between zones of high earthquake or volcanic activity and tectonic plate boundaries	Does not show understanding of the relationship between zones of high earthquake or volcanic activity and tectonic plates
15 The student knows and understands how natural hazards affect human activities	Uses higher-order thinking and a variety of media to identify areas at high risk for a volcanic or seismic disaster, taking into account population density	Uses higher-order thinking to identify areas at high risk for a volcanic or seismic disaster, taking into account population density	Identifies areas at risk but does not take into account population density	Has difficulty identifying areas at risk

This is a four-point rubric based on the National Standards for Geographic Education. The mastery level meets the target objective for grades 5–8.

High school

Standard	Exemplary	Mastery	Introductory	Does not meet requirements
1 The student knows and understands how to use geographic representations and tools to analyze and explain geographic problems	Analyzes GIS data to create a paper map showing zones of high, medium, and low risk for volcanic and earthquake activities and provides a detailed explanation	Analyzes GIS data to create a paper map showing zones of high, medium, and low risk for volcanic and seismic activities and provides some explanation	Analyzes GIS data to create a paper map showing zones of high, medium, and low risk for volcanic and seismic activities	Does not correctly identify risk zones
4 The student understands the changing physical and human characteristics of places	Correctly identifies subduction zones and explains in detail through words and images how they affect three Pacific Rim physical features	Correctly identifies subduction zones and explains how they affect three Pacific Rim physical features	Correctly identifies subduction zones and explains how they affect one or two Pacific Rim physical features	Correctly identifies a subduction zone but does not explain its effect on a physical feature
7 The student knows and understands the spatial variation in the consequences of physical processes across Earth's surface	Clearly explains the correlation of the spatial distribution of volcanoes and earthquakes to plate boundaries and other physical features	Explains the correlation of the spatial distribution of volcanoes and earthquakes to plate boundaries	Identifies the correlation between volcanoes, earthquakes, and plate boundaries but has difficulty explaining it	Identifies the similarities in the spatial distributions of volcanoes and earthquakes but not the correlation to plate boundaries
15 The student understands how humans perceive and react to natural disasters	Uses analytical thinking and a variety of media to formulate a thorough emergency action plan for a city at high risk	Uses analytical thinking to formulate an emergency action plan for a city at high risk	Formulates an emergency action plan that lacks detail and evidence of higher-order thinking	Identifies a city at high risk but does not formulate an emergency action plan

This is a four-point rubric based on the National Standards for Geographic Education. The mastery level meets the target objective for grades 9–12.

51

MODULE 3

Climate

Lesson 1: Running hot and cold: A global perspective

Students will explore characteristics of the Earth's tropical, temperate, and polar zones by analyzing monthly and annual temperature patterns in cities around the world. In the course of their investigation, students will observe temperature patterns associated with changes in latitude as well as differences caused by factors such as elevation and proximity to the ocean.

Lesson 2: Seasonal differences: A regional investigation of South Asia

Students will observe patterns of monsoon rainfall in South Asia and analyze the relationship of those patterns to the region's physical features. The consequences of monsoon season on human life will be explored by studying South Asian agricultural practices and patterns of population distribution.

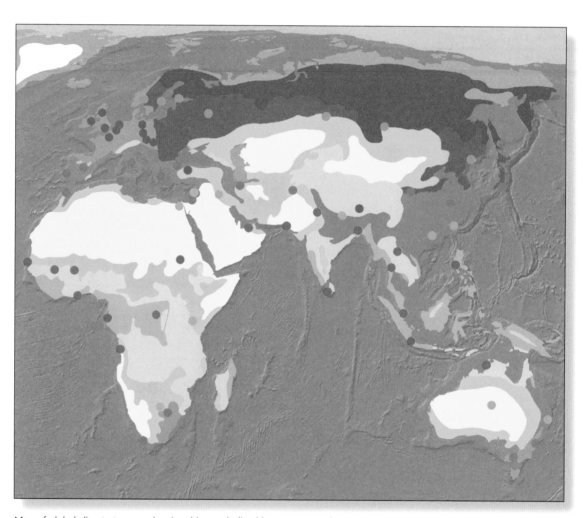

Map of global climate types and major cities symbolized by average yearly temperature.

Running hot and cold

A global perspective

Lesson overview

Students will explore characteristics of the earth's tropical, temperate, and polar zones by analyzing monthly and annual temperature patterns in cities around the world. In the course of their investigation, students will observe temperature patterns associated with changes in latitude as well as differences caused by factors such as elevation and proximity to the ocean.

Estimated time

Two to three 45-minute class periods

Materials

The student worksheet files can be found on the Data and Resources CD. Install the teacher resources folder on your computer to access them.

Location: OurWorld_teacher\Module3\Lesson1
- Student PDF: M3L1_student.pdf
- Student answer sheet: M3L1_student_answer_sheet.doc
- Student supplements: M3L1_supplement.pdf
- Student assessments: M3L1_assessment.pdf

Objectives

After completing this lesson, a student is able to do the following:

- Locate tropical, temperate, and polar zones
- Describe the characteristic yearly and monthly temperature patterns in those zones
- Describe the influence of latitude, elevation, and proximity to the ocean on yearly temperature patterns
- Compare and explain monthly temperature patterns in the Northern and Southern hemispheres.

GIS tools and functions

Icon	Function
A ▾	Change the font color
🔍	Zoom in on the map
A	Add new text to the map
●	Zoom to the full extent of the map
▸	Select and move text on the map
i	Identify features to learn more about them
✚	Add layers to the map
▨	Select features on the map
▨	Clear selected features
✋	Move the map to bring a different portion of it into view
🔍	Find features on the map
💾	Save changes to the map document

- Turn layers on and off
- Set the font, size, and style for feature labels
- Turn feature labels on and off
- Set the font, color, size, and style for text
- Add text to the map
- Activate a data frame
- Display a graph
- Set the selectable layers
- Determine latitude and longitude of map features
- Open an attribute table
- Select records in a table
- Sort records in a table
- Freeze a field in a table
- Clear selected records in a table

National Geography Standards

Standard	Middle school	High school
1 How to use maps and other geographic representations, tools, and technologies to acquire, process, and report information from a spatial perspective	The student understands how to use maps, charts, and databases to analyze spatial distributions and patterns	The student understands how to use technologies to represent and interpret Earth's physical and human systems
5 That people create regions to interpret Earth's complexity	The student understands the elements and types of regions	The student understands the structure of regional systems
7 The physical processes that shape the patterns of Earth's surface	The student understands how Earth–Sun relationships affect physical processes and patterns on Earth	The student understands spatial variation in the consequences of physical processes across Earth's surface

M ● ● ● ○ ○ ○ ○

L ● ○

Teaching the lesson

Introducing the lesson

Begin the lesson by asking students to name places that they believe to be the coldest and hottest on the planet. Briefly compare their choices and the reasoning behind them. Using the "Hot and cold cities" Supplement, have the students work in pairs or small groups to identify the three hottest cities in July and three coldest cities in January. At the end of five minutes, each group should share their lists with the rest of the class. Use the blackboard or an overhead projector to tally the cities mentioned as each group reports. Based on the tally, circle the cities that were listed most often. Explain that they are going to do an activity that will explore temperature patterns in cities around the world. As they complete the activity, they will have an opportunity to check their own answers on this handout and reconsider them in view of what they learn.

Before beginning the computer activity, engage students in a discussion about the cities that are circled on the list.

* Why do you think this city is one of the coldest or hottest?
* What countries are these cities located in?
* Has anyone ever visited one of these cities?

Student activity

We recommend that you complete the activity yourself before presenting the lesson in class. Doing so will allow you to modify the activity to accommodate the specific needs of your students. If they will not be working on individual computers, be sure to explain any necessary modifications.

Explain to students that in this activity they will use GIS to analyze yearly and monthly temperature patterns in cities around the world. They will identify global and regional temperature variations and speculate on possible reasons for the patterns observed.

Make sure you inform students how to rename their map document and where to save it for question Q8 (at the end of step 4). This is a good place to end a work session. Another good break point is before students begin step 9.

The following are things to look for while the students are working on this activity:

* Are students using a variety of GIS tools?
* Are students answering the questions?
* Are students experiencing any difficulty navigating between windows in the map document?

Concluding the lesson

When the class has finished the activity, lead a discussion that summarizes the conclusions the students reached. Be sure to address latitude in the Northern Hemisphere, latitude in the Southern Hemisphere, proximity to ocean, and elevation as factors that influence temperature. After students have had an opportunity to share their conclusions, discuss the similarities and differences among the ideas presented. Allow students to question each other and clarify confusing or contradictory statements. Develop a consensus about how each factor influences temperature.

Assessment. Students will draw conclusions about the factors that influence temperature patterns. They will write an essay offering data and examples from the activity that support these statements. Middle school students need to make a paper map or a GIS-generated map that illustrates the points of their essay. High school students need to use ArcMap only.

Extending the lesson

Challenge students to try the following:

- Collect additional temperature data for cities in one region. Use that data to create a map document of that region and a regional temperature profile.
- Investigate the phenomenon of global warming. Use the Internet to collect recent monthly temperature data for one or more of the cities included in the map document. Compare actual recorded temperatures to average monthly temperatures to see if current temperatures are warmer than average. Compare changes in one region with global changes to see if there are differences.
- Collect rainfall data for the cities included in the map document. Use this data in combination with the temperature data to create an ArcMap layout that illustrates typical climate patterns.

See the "Resources by Module" section of this book's Web site—www.esri.com/ourworldgiseducation— for print, media, and Internet resources on the topics of climate and global temperatures.

Answer key

Step 2: Observe annual world temperature patterns

Q1. Write three observations about the pattern of temperatures displayed on the map. Your observations should describe regions of the world, not specific countries or cities. Student answers will vary. Possible observations include the following:

The warmest temperatures are clustered halfway between the North and South poles.

Temperatures get steadily colder as you go from the equator toward the North Pole.

There are many cities with cold temperatures in the Northern Hemisphere, but none in the Southern Hemisphere.

Step 3: Label latitude zones

Q2. Complete the following table.

Zone	Typical temperature range (°C)	Typical city	Atypical cities
Tropical	19 to 29	Any of the 28 cities colored red or orange that are between the Tropic of Capricorn and the Tropic of Cancer	Quito and La Paz
North Temperate	-1 to 18	Any of the cities colored pink, blue, or green that are between the Tropic of Cancer and the Arctic Circle	Students should look for cities that seem to differ from the others around them or from other cities at the same latitude, such as Lhasa and Ankara.
South Temperate	13 to 23	Any of the cities colored green or orange that are between the Tropic of Capricorn and the Antarctic Circle	None

Q3. Why do you think there aren't any major cities in the North or South Polar Zones? The temperatures are probably too cold to support major cities.

Q4. How is the North Temperate Zone different from the South Temperate Zone?
Answers will vary and may include the following:
There is more land area in the North Temperate Zone.
There are cities with average temperatures below 13°C in the North Temperate Zone, but none in the South Temperate Zone.

Step 4: Observe climate distribution

Q5. Complete the following table.

Zone(s)	Characteristic climate(s)
Tropical	Tropical Wet, Tropical Dry Some areas of Arid, Semiarid, Humid Subtropical, Highlands
Temperate	Humid Subtropical, Humid Continental, Marine, Mediterranean, Subarctic Some areas of Arid, Semiarid, Highlands, Tundra
Polar	Subarctic, Tundra, Ice Cap

Q6. Which zone has the greatest number of climates? Temperate Zones (or North Temperate Zone)

Q7. Give an example of a city in each of the following climate zones:
Answers will vary and may include the following:
Arid Khartoum
Tropical Wet Kisangani
Tropical Dry Bamako
Humid Subtropical Atlanta
Mediterranean Roma (Rome)
Marine Paris
Humid Continental Warsaw
Subarctic Irkutsk
Highland Lhasa

Q8. Record the new name of the map document and its new location.
Document Example: ABC_Mod3Les1.mxd
Location Example: C:\Student\ABC

Step 5: Observe monthly temperature patterns in the Northern Hemisphere

Q9. What does the graph show now? Average monthly temperatures in Miami

Q10. What city is highlighted on the map? Miami

Q11. What does the graph show now? Average monthly temperatures in Miami and Boston

Q12. What city or cities are highlighted on the map? Miami and Boston

Q13. Complete the following table. Student answers for temperature may vary slightly depending on how they interpret the graph. Answers below are the actual attribute values.

City	Coldest months	Lowest temperature (°C)	Hottest month(s)	Highest temperature (°C)	Temperature range over 12 months (°C)
Boston	January, February	-2	July	23	25
Miami	January, February	20	July, August	28	8

Q14. What is the name of the city? Quebec

Q15. How does its monthly temperature pattern differ from Boston's? The overall pattern is the same, but winter temperatures are colder and summer temperatures are slightly cooler. The annual temperature range is greater: 31°C. Summer is slightly shorter and winter slightly longer in the more northern city.

Q16. What is the name of the city? Kingston

Q17. How does its monthly temperature pattern differ from Miami's? Kingston has a smaller temperature range (2°C) than Miami. Both cities are warm year-round, but Miami shows more seasonal variation. They have identical high temperatures, but Kingston's lows are not as cool as those in Miami.

Q18. Complete the following table for each of the cities displayed in the graph. Answers for latitude will vary slightly.

City	Latitude	Coldest month(s)	Lowest temperature (°C)	Hottest month(s)	Highest temperature (°C)	Temperature range over 12 months
Quebec	46.9	January	-12	July	19	31
Boston	42.4	January, February	-2	July	23	25
Miami	25.8	January, February	20	July, August	28	8
Kingston	18.1	December to March	26	June to September	28	2

Q19. Based on the information displayed in the graph, the map, and the table in Q18, formulate a hypothesis about how the monthly temperature patterns change as latitude increases. Answers will vary but should include the following points:
As latitude increases, the range of temperatures over the year increases.
The lower latitudes have less seasonal variation and tend to be warm year-round.
Temperatures get steadily colder as latitude increases.
January and February are among the coldest months, and July and August are among the hottest months. (Differences between the Northern and Southern hemispheres will be addressed later in the lesson.)

Module 3: Lesson 1

Step 6: Test your hypothesis

Q20. Complete the following table. Student answers will vary.

City	Latitude (°)
Stockholm	59
Warsaw	52
Budapest	48
Athinai (Athens)	38

Q21. Does the data for the cities you selected confirm or dispute your hypothesis in Q19? Explain. Answers will vary depending on the hypotheses. Like the pattern observed in North America, these cities get warmer as you move south. Coldest temperatures occur in December and January, while warmest temperatures occur in June through August. Unlike in North America, the annual range of temperatures varies little among these cities.

Step 7: Analyze temperature patterns in the Southern Hemisphere

Q22. Complete the following table.

City	Latitude (°)	Coldest month(s)	Lowest temperature (°C)	Hottest month(s)	Highest temperature (°C)	Temperature range over 12 months
Darwin	-13	July	25	April, October to December	29	4
Brisbane	-27	July	15	January, February	25	10
Sydney	-34	July	12	December to February	22	10
Melbourne	-38	June, July	9	February	20	11

Q23. Compare the monthly temperature patterns in the Southern Hemisphere to those in the Northern Hemisphere. Patterns in the Southern Hemisphere mirror those in the Northern Hemisphere. The coldest winter temperatures are not as low as for the Northern Hemisphere because none of the cities has a latitude higher than -38°. The major difference is that the warmest and coldest months are reversed.

Q24. Formulate a hypothesis about the relationship between monthly temperature patterns and increases in latitude in the Southern Hemisphere. Answers will vary.

Step 8: Test your hypothesis for the Southern Hemisphere

Q25. Complete the following table. Possible answers are listed in the table. Libreville and Nairobi are also correct cities.

City	Latitude (°)
Cape Town	-34
Johannesburg	-26
Gabarone	-25
Luanda	-9

Q26. Does the data for the cities you selected confirm or dispute your hypothesis about how latitude affects monthly temperature patterns in the Southern Hemisphere? Explain. Answers will vary. The seasonal pattern is similar to that observed in Australia.

Step 9: Investigate the ocean's influence on temperature

Q27. In which Canadian city would you experience the coldest winter temperatures? Winnipeg

Q28. In which Canadian city would you experience the warmest winter temperatures? Vancouver

Q29. Looking at the map, why do you think the warmest city has winter temperatures that are so much warmer than the others? Of the selected Canadian cities, Vancouver is the only one located on the coast. The proximity to the ocean has a steadying effect on the air temperature in Vancouver throughout the year. Therefore, the fluctuation between summer and winter temperatures is not as large as with inland cities at the same latitude.

Q30. Complete the following table.

City	Latitude (°)
London	52
Amsterdam	52
Berlin	53
Warsaw	52
Kiev	50

Q31. What do these cities have in common in terms of their locations on the earth? All the cities are in the Northern Hemisphere, on the continent of Europe, and at approximately 50° north latitude.

Q32. Which two cities have the mildest temperatures? London and Amsterdam

Q33. What happens to the winter temperatures as you move from London to Kiev? Winter temperatures get steadily colder as you move east and inland.

63

Q34. Why do you think some cities have milder temperatures than the others? The warmest cities are near the ocean (London on an island, Amsterdam on the coast).

Q35. Based on your observations for Canada (Q27–Q29) and Western Europe, formulate a hypothesis about the influence of proximity to the ocean (or distance from it) on patterns of temperature. Answers will vary but should note that cities closest to the ocean have milder temperatures than cities at similar latitudes located inland.

Step 10: Investigate the impact of elevation on temperature patterns

Q36. Complete the following table.

City	Latitude (°)
Kisangani	1
Libreville	0
Quito	-1
Singapore	1

Q37. What do these cities have in common in terms of their locations on the earth? All are located very close to the equator.

Q38. What temperature pattern do these four cities have in common? All four cities show very little range in monthly temperatures throughout the year (3° or less).

Q39. How is Quito different from the other three? Its temperatures are significantly cooler than the temperatures of the other three cities.

Q40. Since all these cities are located on or very near the equator, what other factor could explain the difference in their temperature patterns? Answers will vary. Students should not predict that Quito's close proximity to the ocean causes its cooler temperatures, because they just learned that proximity to the ocean causes milder temperatures.

Q41. Analyze the selected records and complete the following table.

City	Elevation (meters)
Kisangani	415
Libreville	10
Quito	2812
Singapore	32

Q42. Based on your observation of temperatures along the equator and the information in the table in Q41, formulate a hypothesis about the influence of elevation on patterns of temperature. Answers will vary but should note that cities at significantly higher elevations have cooler temperatures than other cities at a similar latitude.

Step 11: Revisit your initial ideas

Q43. Rank the 13 cities from coldest to hottest according to their average January temperatures.

1. Irkutsk	8. Quito
2. Minneapolis	9. Wellington
3. Helsinki	10. Miami
4. Lhasa	11. Khartoum
5. Vancouver	12. Buenos Aires
6. London	13. Singapore
7. Tunis	

Q44. Rank the 13 cities from hottest to coldest according to their average July temperatures.

1. Khartoum	8. Helsinki
2. Miami	9. Vancouver
3. Singapore	10. Lhasa
4. Tunis	11. Quito
5. Minneapolis	12. Buenos Aires
6. London	13. Wellington
7. Irkutsk	

Q45. Put a check mark next to the answers in Q43 and Q44 that you predicted correctly. Answers will vary.

Assessment rubrics

Middle school

Standard	Exemplary	Mastery	Introductory	Does not meet requirements
1 The student understands how to use maps, graphs, and databases to analyze spatial distributions and patterns	Uses GIS to analyze various aspects of climate such as monthly and annual temperature and precipitation and identifies cities with specific climate characteristics; creates a detailed map that illustrates the points highlighted in the essay	Uses GIS to analyze various aspects of climate such as monthly and annual temperature and precipitation and identifies cities with specific climate characteristics; creates a map that illustrates points highlighted in the essay	With some assistance, can use GIS to analyze various aspects of climate such as monthly and annual temperature and precipitation; correctly identifies some cities with specific climate characteristics; creates a map that illustrates some of the points highlighted in the essay	Has difficulty using GIS to analyze various aspects of climate and identifying cities with specific climate characteristics; creates a map but has difficulty illustrating the points highlighted in the essay
5 The student understands the elements and types of regions	Writes an essay and creates a map that shows clear understanding of the climate patterns for various zones of the Earth, including the Tropics and the North and South Temperate zones	Writes an essay and creates a map that shows an understanding of the climate patterns for various zones of the Earth, including the Tropics and the North and South Temperate zones	Writes an essay and creates a map that shows some understanding of climate patterns but does not clearly define differences between different zones, or identifies characteristics of only some zones	Writes an essay and creates a map that shows limited understanding of climate patterns and cannot identify differences between various zones
7 The student understands how Earth–sun relationships affect physical processes and patterns on Earth	Identifies key reasons and provides clear examples of why cities experience variations in climate patterns at different latitudes, in different hemispheres, at different elevations, and at different distances from the ocean	Identifies key reasons why cities experience variations in climate patterns at different latitudes, in different hemispheres, at different elevations, and at different distances from the ocean	Identifies some key reasons why cities experience variations in climate patterns at different latitudes, in different hemispheres, at different elevations, or at different distances from the ocean	Identifies one or two reasons why cities experience variations in climate patterns at different latitudes, in different hemispheres, at different elevations, or at different distances from the ocean

This is a four-point rubric based on the National Standards for Geographic Education. The mastery level meets the target objective for grades 5–8.

Module 3: Lesson 1

High school

Standard	Exemplary	Mastery	Introductory	Does not meet requirements
1 The student understands how to use technologies to represent and interpret Earth's physical and human systems	Uses GIS to analyze various aspects of climate such as monthly and annual temperature, precipitation, elevation, and proximity to the ocean and identifies cities with specific climate characteristics; uses GIS to create a detailed map that illustrates the points highlighted in the essay	Uses GIS to analyze various aspects of climate such as monthly and annual temperature, precipitation, elevation, and proximity to the ocean and identifies cities with specific climate characteristics; uses GIS to create a map that illustrates points highlighted in the essay	With some assistance, can use GIS to analyze various aspects of climate such as monthly and annual temperature and precipitation; correctly identifies some cities with specific climate characteristics; uses GIS to create a map that illustrates some of the points highlighted in the essay	Has difficulty using GIS to analyze various aspects of climate and identifying cities with specific climate characteristics; uses GIS to create a map but has difficulty illustrating the points highlighted in the essay
5 The student understands the structure of regional systems	Writes an essay and creates a GIS-generated map that shows clear understanding of the climate patterns for various zones of the Earth, including the Tropics and the North and South Temperate zones	Writes an essay and creates a GIS-generated map that shows an understanding of the climate patterns for various zones of the Earth, including the Tropics and the North and South Temperate zones	Writes an essay and creates a GIS-generated map that shows some understanding of climate patterns but does not clearly define differences between different zones, or identifies characteristics of only some zones	Writes an essay and creates a GIS-generated map that shows limited understanding of climate patterns and cannot identify differences between various zones
7 The student understands spatial variation in the consequences of physical processes across Earth's surface	Identifies key reasons and provides clear examples of why cities experience variations in climate patterns at different latitudes, in different hemispheres, at different elevations, and at different distances from the ocean	Identifies key reasons why cities experience variations in climate patterns at different latitudes, in different hemispheres, at different elevations, and at different distances from the ocean	Identifies some key reasons why cities experience variations in climate patterns at different latitudes, in different hemispheres, at different elevations, or at different distances from the ocean	Identifies one or two reasons why cities experience variations in climate patterns at different latitudes, in different hemispheres, at different elevations, or at different distances from the ocean

This is a four-point rubric based on the National Standards for Geographic Education. The mastery level meets the target objective for grades 9–12.

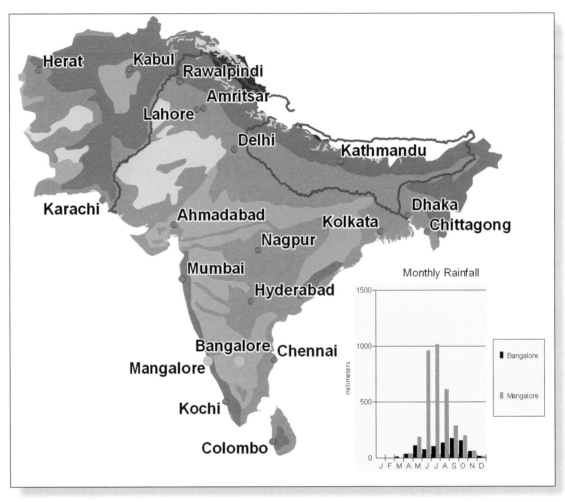

The influence of physical features on weather patterns is explored using maps and graphs.

Module 3: Lesson 2

Seasonal differences
A regional investigation of South Asia

Lesson overview
Students will observe patterns of monsoon rainfall in South Asia and analyze the relationship of those patterns to the region's physical features. The consequences of monsoon season on human life will be explored by studying South Asian agricultural practices and patterns of population distribution.

Estimated time
Two to three 45-minute class periods

Materials
The student worksheet files can be found on the Data and Resources CD. Install the teacher resources folder on your computer to access them.

Location: OurWorld_teacher\Module3\Lesson2
• Student PDF: M3L2_student.pdf
• Student answer sheet: M3L2_student_answer_sheet.doc
• Student assessments: M3L2_assessment.pdf

Additional materials
• Four large pieces of butcher paper
• Four or more markers

Objectives
After completing this lesson, a student is able to do the following:

• Describe the patterns of monsoon rainfall in South Asia
• Explain the influence of landforms on patterns of precipitation
• Describe the impact of climate and physical features on agriculture and population density

GIS tools and functions

 Select features on a map

 Measure distance on a map

 Clear selected features

 Add layers to a map

- Display graphs stored with the map document
- Analyze graphs in relation to a map
- Understand the relationship between a graph and a map
- Set selectable layers
- Rearrange layers in the table of contents
- Turn layers on and off

National Geography Standards

Geography standard	Middle school	High school
1 How to use maps and other geographic representations, tools, and technologies to acquire, process, and report information from a spatial perspective	The student understands how to use maps, graphs, and databases to analyze spatial distributions and patterns	The student understands how to use technologies to represent and interpret Earth's physical and human systems
4 The physical and human characteristics of places	The student understands how physical processes shape places	The student understands the changing human and physical characteristics of places
15 How physical systems affect human systems	The student understands how variations within the physical environment produce spatial patterns that affect human adaptation	The student understands how the characteristics of different physical environments provide opportunities for or place constraints on human activities

Teaching the lesson

Introducing the lesson

Tell your students that they are going to explore seasonal differences in South Asia. They may be surprised to learn that students in South Asia would probably describe their year in terms of three seasons rather than four. Engage them in a discussion of local and personal perceptions and assumptions about seasons. Tack up four large pieces of butcher paper and have them list images, descriptions, and memories relating to each season.

- How does the physical environment change from season to season?
- How are those changes reflected in activities, foods, and clothing?
- To what extent do seasonal changes in their environment affect their day-to-day lives?

M ● ● ● ○ ○ ○ ○

L ● ● ●

Student activity

We recommend that you complete the activity yourself before presenting the lesson in class. Doing so will allow you to modify the activity to accommodate the specific needs of your students. If they will not be working on individual computers, be sure to explain any necessary modifications.

Explain to students that in this activity they will use GIS to observe and analyze the variable patterns of rainfall in South Asia that result from the region's seasonal monsoon winds. In South Asia it is rainfall, rather than temperature, that defines the seasons. Students will explore the relationship between South Asia's monsoon rains and its physical features and analyze the climate's impact on agriculture and population.

The following are things to look for while the students are working on this activity:

- Are students using a variety of GIS tools?
- Are students answering the questions?
- Are students experiencing any difficulty navigating between windows in the map document?

Concluding the lesson

Use a projection device to display the **Region3.mxd** map document in the classroom. As a group, compare student observations and conclusions from the lesson. Students can take turns being the "driver" on the computer to highlight patterns and relationships that are identified by members of the class. Focus on the following concepts about South Asia's monsoon climate in your discussion:

- Rainfall is limited to one season of the year in South Asia except in the desert west, where little rain falls at all. (This would be an excellent point at which to elaborate on the seasonal shift in monsoon winds that produces the patterns of rainfall students observed in the map document.)
- Typically, the rainy season lasts from June through September, although the actual length of the season and amounts of rainfall vary across the subcontinent. (Be sure to note the orographic patterns of precipitation along India's southwest coast and in northeast India on the southern slopes of the Himalayas.)
- Agricultural activities are directly related to patterns of rainfall.
- In general, population density varies with patterns of rainfall. However, the importance of South Asia's rivers as an additional source of water for agriculture is apparent from the high density of population along their paths.

Close the lesson by challenging the students to identify the three seasons in South Asia. In general these seasons are the following:

- The rainy season (approximately June–September)
- The dry lush season after the rains when everything is growing and green (approximately October–January)
- The dry dusty season before the rains come (approximately February–May)

Middle school assessment. Students will assume the role of an American student living for a year in South Asia as an exchange student. They can choose to live in or around Mumbai, Delhi, or Kolkata. They will write a letter to friends back home on January 1, April 1, July 1, and October 1. Their letter will describe seasonal changes in their location and ways that their daily lives and the lives of people around them reflect those changes.

High school assessment. Students will assume the role of an American student traveling for a year in South Asia. They will write a letter to friends back home on January 1, April 1, July 1, and October 1. Each letter will be from a different South Asian city. Their letter will describe seasonal characteristics in each city and ways that their daily lives and the lives of people around them reflect those characteristics.

Extending the lesson

Challenge students to try the following:

* Use the Internet to find rainfall data for South Asian cities in specific years such as 1990, 1995, 2000, and 2005 to see if the average patterns observed in this lesson are relatively consistent or if they vary significantly from year to year.
* Import downloaded data into ArcMap.
* Research South Asian farming methods to find out how activities such as planting and harvesting are coordinated with patterns of rainfall.
* Research the monthly and yearly rainfall patterns in your own location and compare these to the patterns observed in South Asia.

See the "Resources by Module" section of this book's Web site—www.esri.com/ourworldgiseducation— for print, media, and Internet resources on the topics of South Asia and monsoons.

Answer key

Step 2: Observe patterns of rainfall

Q1.   Which month gets the most rainfall in Mumbai? July

Q2.   Which months appear to get little or no rainfall in Mumbai? December–April

Q3.   Approximately how much rainfall does Mumbai get each year (in millimeters)? 2,100

Q4.   Write a sentence summarizing the overall pattern of rainfall in Mumbai in an average year. Mumbai gets more than 2,100 millimeters of rain per year in a concentrated period from June to September.

Q5.   How did this change the map? The new city is selected. It turns blue on the map.

Q6.   How did this change the graphs? Both graphs now show data for the new city.

Q7.   Analyze the graphs and fill in the Mangalore section of the table below. Estimate the rainfall amounts.

City	Months with rainfall > 50 mm	Highest monthly rainfall (mm)	Total annual rainfall (mm)
Mangalore	May–November	1,000	3,500
Mumbai			
Ahmadabad			

Q8.   Complete the rest of the table in Q7. Use estimates.

City	Months with rainfall > 50 mm	Highest monthly rainfall (mm)	Total annual rainfall (mm)
Mangalore	May–November	1,000	3,500
Mumbai	June–October	650	2,000
Ahmadabad	June–September	300	800

Q9.   As you move northward along the subcontinent's west coast, how does the pattern of rainfall change? The rainy season gets shorter. It starts later in the year and ends earlier. The monthly and yearly rainfall totals decline.

Q10. Although the monthly rainfall amounts differ, what similarities do you see among the overall rainfall patterns of these three cities? The rainy season occurs at the same time of year in all three cities; so does the dry season. In each city, July has the highest rainfall total of any month, and the period from December through March is dry.

Step 3: Compare coastal and inland cities

Q11. Complete the table below.

City	Months with rainfall > 50 mm	Highest monthly rainfall (mm)	Total annual rainfall (mm)
Bangalore	May–November	175	900

Q12. How does the rainfall pattern of Bangalore compare with that of Mangalore?
Similarities: The two cities have a rainy season between May and November.
Differences: Mangalore gets approximately four times as much rain in a year as Bangalore.

Q13. What is the distance between the two cities? Approximately 270 kilometers

Q14. How can the data in the Physical Features layer help you explain the differences between patterns of rainfall in inland Bangalore and coastal Mangalore? Mangalore is on the coast, while Bangalore is on the interior (Deccan) plateau. A narrow coastal mountain range (the Western Ghats) separates the two cities. The significant difference in total and monthly rainfall results from the orographic effect produced by the Western Ghats. Moist monsoon winds are forced to rise to go over these mountains as they come ashore. Condensing in the cooler upper atmosphere, most of the monsoon's moisture falls on the windward side of the mountains, leaving the inland side much drier.

Step 4: Compare eastern and western South Asian cities

Q15. Analyze the graphs and complete the table below.

City	Months with rainfall $\geq$ 20 mm	Highest monthly rainfall (mm)	Total annual rainfall (mm)
Kabul	December–May	70	280
Herat	December–April	55	245

Q16. Describe the pattern of rainfall in these two cities. Both of these cities are extremely dry. What little rainfall they do receive falls in the early months of the year.

Q17. How do you think Afghanistan's rainfall pattern affects the way of life in that country? There is not enough rainfall to support crop farming. They have to rely on nomadic herding and extraction of natural resources.

Q18. Analyze the graphs and complete the table below.

City	Months with rainfall $\geq$ 20 mm	Highest monthly rainfall (mm)	Total annual rainfall (mm)
Kolkata	February–November	338	1,633
Dhaka	February–November	399	1,996

Q19. Describe the pattern of rainfall in these two cities. These two cities have significant annual rainfall totals and a distinct rainy season that lasts longer than the rainy season on the southwest coast. The majority of the rain falls between May and October.

Q20. What is happening to the patterns of rainfall as you move from west to east across South Asia? The amount of annual rainfall increases as you move eastward, and the rainy season gets longer.

Step 5: Observe yearly precipitation

Q21. Which regions of South Asia get the least rainfall? The northwest (Afghanistan and Pakistan)

Q22. Which regions of South Asia get the most rainfall? The southwest coast and the northeast

Q23. In Q20 you were comparing Herat, Delhi, Kolkata, and Dhaka. Does the map of yearly rainfall that is on your screen now reflect the observation you made at that time? Explain. Answers will vary, but essentially, students should observe that precipitation does increase as you move from west to east across South Asia.

Q24. What relationships do you see between South Asia's patterns of yearly rainfall and its physical features? The region's heaviest rainfall is on the windward side of the Western Ghats and the Himalayas. Orographic lift is responsible for these areas of heavy rainfall. Cities on the Deccan Plateau, in the interior of the subcontinent, get significantly less rainfall because they lie in the rain shadow of the mountains.

Step 6: Explore the monsoon's impact on agriculture and population density

Q25. Which regions or countries of South Asia are suitable for crop farming and which are not? Explain. Answers will vary. The western section of South Asia (Afghanistan, Pakistan, and western India) does not get enough rainfall to support crop farming. Additionally, much of Afghanistan and Pakistan is in the mountains, making crop farming unlikely there. Most of the remainder of the subcontinent is suitable for crop farming because it gets sufficient rainfall and is either a plain or plateau.

Q26. In which regions of South Asia do you expect to see the lowest population density? Explain. Answers will vary. Students should expect the dry mountainous west to have the lowest population density because the region cannot produce enough food to support a large population.

Q27. In which regions of South Asia do you expect to see the highest population density? Explain. Answers will vary. Students should recognize the importance of rivers to agriculture (alluvial floodplain, fertile deltas, and a steady source of water) and predict a high population density there.

Q28. Does the Agriculture layer reflect the predictions you made in step Q25? Explain. Answers will vary depending on the answers in Q25. However, the data does illustrate lack of crop farming in the dry mountainous regions.

Q29. Why are grazing, herding, and oasis agriculture the major activities in Afghanistan? Mountainous terrain and scarce rainfall make these the only viable economic activities for most people.

Q30. What do you know about rice cultivation that would help explain its distribution on the agriculture map? Students familiar with rice cultivation will note that this is a crop that is often grown in flooded fields (wet rice cultivation) and requires a lot of water. Therefore, it makes sense that rice is cultivated in areas with significant rainfall.

Q31. Is there any aspect of the agriculture map that surprised you? Explain. Answers will vary. Some students may be surprised about the agricultural activity in Pakistan, since the area is so dry.

Q32. Does the Population Density layer reflect the population predictions you made in Q26 and Q27? Explain. Answers will vary depending on the predictions in Q26 and Q27.

Q33. Why is Afghanistan's population density so low? Its mountainous terrain and lack of rain make this an area that cannot support a large population.

Q34. Since most of Pakistan gets little to no rainfall, how do you explain the areas of high population density in that country? The Indus River provides a rich alluvial floodplain and a year-round supply of water for irrigation.

Q35. What is the relationship between population density and patterns of precipitation in South Asia? Overall population density is highest where rainfall amounts are conducive to agriculture. The notable variation to this pattern is the high population density along the rivers—particularly in the west. The rich soil and dependable source of water on the Indo-Gangetic Plain enable agriculture to support dense populations in spite of insufficient rainfall in some areas or at some times of year.

Q36. What is the relationship between population density and physical features in South Asia? Population density is lowest in mountainous areas of Afghanistan and Pakistan and highest on the Indo-Gangetic Plain.

Assessment rubrics

Middle school

Standard	Exemplary	Mastery	Introductory	Does not meet requirements
1 The student understands how to use maps, graphs, and databases to analyze spatial distributions and patterns	Uses GIS as a tool to analyze the patterns of monsoon rains in South Asia through mapping and creating original charts based on the data provided; accurately depicts the seasonal weather conditions in a given city	Uses GIS as a tool to analyze the patterns of monsoon rains in South Asia through mapping and viewing charts; accurately depicts most of the seasonal weather conditions in a given city	Identifies patterns of precipitation on maps and charts using GIS; accurately depicts some of the seasonal weather conditions for a given city	Has difficulty identifying patterns of precipitation using maps and charts in GIS; has difficulty depicting any seasonal weather conditions for the given city
4 The student understands how physical processes shape places	Shows understanding of the effects of seasonal rains and other climate factors on the characteristics of many South Asian cities throughout an entire year; creatively incorporates this into each letter	Shows understanding of the effects of seasonal rains and other climate factors on the characteristics of one particular South Asian city throughout an entire year; incorporates this into each letter	Shows limited understanding of the effects of seasonal rains on the characteristics of one particular South Asian city throughout an entire year; incorporates some of this information into each letter	Attempts to describe seasonal climate changes in the region of South Asia; has difficulty incorporating this information into each letter
15 The student understands how variations within the physical environment produce spatial patterns that affect human adaptation	Uses great detail and specific examples to illustrate the impact of seasonal climate changes on the daily life of people in a variety of South Asian cities throughout an entire year; creatively incorporates this into each letter	Clearly illustrates the impact of seasonal climate changes on the daily life of people in a particular South Asian city throughout an entire year; incorporates this into each letter	Shows limited understanding of the impact of seasonal climate changes on the daily life of people in a particular South Asian city throughout an entire year; incorporates some of this information into each letter	Attempts to describe the impact of seasonal climate changes on the daily life of people in South Asia; has difficulty incorporating this information into each letter

This is a four-point rubric based on the National Standards for Geographic Education. The mastery level meets the target objective for grades 5–8.

High school

Standard	Exemplary	Mastery	Introductory	Does not meet requirements
1 The student understands how to use technologies to represent and interpret Earth's physical and human systems	Uses GIS as a tool to analyze the patterns of monsoon rains in South Asia through mapping and creating original charts based on the data provided; accurately depicts the seasonal weather conditions in the four cities	Uses GIS as a tool to analyze the patterns of monsoon rains in South Asia through mapping and viewing charts; accurately depicts most of the seasonal weather conditions in the four cities	Identifies patterns of precipitation on maps and charts using GIS; accurately depicts some of the seasonal weather conditions for some of the cities	Has difficulty identifying patterns of precipitation using maps and charts in GIS; has difficulty depicting the seasonal weather conditions for the cities
4 The student understands the changing human and physical characteristics of places	Shows understanding of the effects of seasonal rains and other climate factors on the characteristics of four South Asian cities; describes the seasonal changes for an entire year for each city; creatively incorporates this into each letter	Shows understanding of the effects of seasonal rains and other climate factors on the characteristics of four different South Asian cities throughout an entire year (one city for each season); incorporates this into each letter	Shows limited understanding of the effects of seasonal rains on the characteristics of one or two South Asian cities throughout an entire year; incorporates some of this information into each letter	Attempts to describe seasonal climate changes in the region of South Asia; has difficulty incorporating this information into each letter
15 The student understands how the characteristics of different physical environments provide opportunities for or place constraints on human activities	Uses great detail and specific examples to illustrate the impact of seasonal climate changes on the daily life of people in four South Asian cities; describes the seasonal changes for an entire year for each city; creatively incorporates this into each letter	Clearly illustrates the impact of seasonal climate changes on the daily life of people in four different South Asian cities throughout an entire year (one city for each season); incorporates this into each letter	Shows limited understanding of the impact of seasonal climate changes on the daily life of people in one or two South Asian cities throughout an entire year; incorporates some of this information into each letter	Attempts to describe the impact of seasonal climate changes on the daily life of people in South Asia; has difficulty incorporating this information into each letter

This is a four-point rubric based on the National Standards for Geographic Education. The mastery level meets the target objective for grades 9–12.

MODULE 4

Populations

Lesson 1: The march of time: A global perspective

Students will analyze the locations and populations of the world's largest cities from the year 100 CE through 2005 CE. They will describe spatial patterns of growth and change among the world's largest urban centers during the past two thousand years and speculate on reasons for the patterns they observe.

Lesson 2: Growing pains: A regional investigation of Europe and Africa

Students will compare the processes and implications of population growth in one of the world's fastest growing regions, sub-Saharan Africa, and the slowest growing region, Europe. Through the analysis of standard-of-living indicators in these two regions, students will explore some of the social and economic implications of rapid population growth.

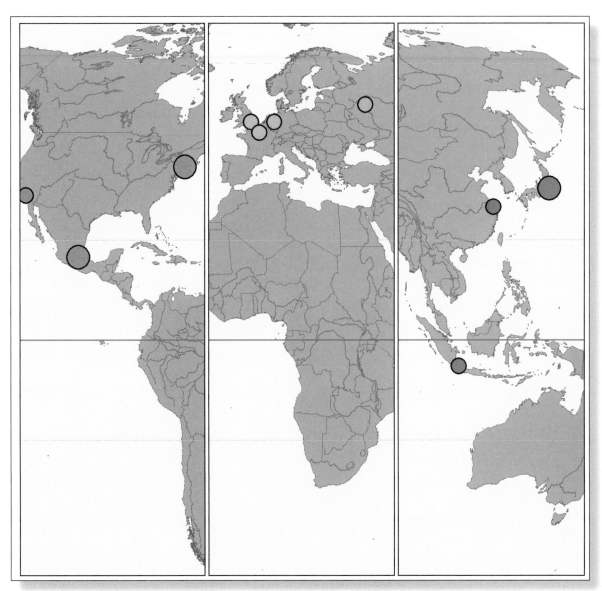

The location and size of global population centers shifts over time (from maps of 2000, 1950, 2005, left to right).

The march of time

A global perspective

Lesson overview

Students will analyze the locations and populations of the world's largest cities from the year 100 CE through 2005 CE, describe spatial patterns of growth and change among the world's largest urban centers during the past two thousand years, and speculate on possible reasons for the patterns they observe.

Estimated time

Two 45-minute class periods

Materials

The student worksheet files can be found on the Data and Resources CD. Install the teacher resources folder on your computer to access them.

Location: OurWorld_teacher\Module4\Lesson1
- Student PDF: M4L1_student.pdf
- Student answer sheet: M4L1_student_answer_sheet.doc
- Student assessments: M4L1_assessment.pdf
- Lesson transparency: M4L1_transparency.pdf

Objectives

After completing this lesson, a student is able to do the following:

- Describe the locations and sizes of the world's largest cities over time
- Identify historical events and periods that influenced the locations of cities throughout history
- Explain the ever-increasing pattern of growth among the world's urban populations in the past two thousand years
- Define agglomeration and how that differs from city proper

GIS tools and functions

Find a specific feature in a layer

Learn more about a selected feature

Clear selected features

Label selected features in a layer

Move or unselect a graphic label

Zoom to the full extent of the map

Add a layer to the map

- Turn layers on and off
- Find a feature and select it
- Zoom to a layer's extent
- Open the attribute table for a layer
- Sort data in descending order
- Select a record in a table
- Clear selected features in all layers
- Zoom to a selected feature

National Geography Standards

Standard	Middle school	High school
1 How to use maps and other geographic representations, tools, and technologies to acquire, process, and report information from a spatial perspective	The student knows and understands how to use maps to analyze spatial distributions and patterns	The student knows and understands how to use geographic representations and tools to analyze and explain geographic problems
12 The processes, patterns, and functions of human settlement	The student knows and understands the spatial patterns of settlement in different regions of the world	The student knows and understands the functions, sizes, and spatial arrangements of urban areas
17 How to apply geography to interpret the past	The student knows and understands how the spatial organization of a society changes over time	The student knows and understands how processes of spatial change affect events and conditions

Teaching the lesson

Introducing the lesson

Begin the lesson by briefly reviewing terms defining urban settlements. For example:

- A **city** is a place where many people live close together.
- An **urbanized area** is a cluster of cities, usually a large central city surrounded by some smaller cities.
- An **urban agglomeration** is where many large and small cities have merged into a very large, extended urban area.

A city's extent can be defined by its legal boundary, as a continuously built-up area, or as a functional area. Explain that in the past, cities had well-defined borders but rapid growth of urban areas in recent history has made it challenging to define their extent and to measure their population. This means that population data collected by different countries or organizations can vary drastically depending on how they define a city's extent.

Next, divide the students into pairs. Challenge each group to name the ten most populated cities in the world today. After five minutes, each group should share their list with the rest of the class. Use the blackboard or an overhead projector to tally the cities mentioned as each group reports. Based on the tally, circle the cities that were listed most often. Tell the class that they are going to do a GIS investigation that will use real data to identify the 10 most populated cities in the world from the last two thousand years.

Finally, engage students in a discussion about the cities circled on their list.

- What do they know about these cities?
- In what countries are these cities located?
- How many people live in these urban centers?
- Has anyone ever visited one of these cities?
- Can they think of any reasons why some cities grow to be so large?

Student activity

We recommend that you complete the activity yourself before presenting the lesson in class. Doing so will allow you to modify the activity to accommodate the specific needs of your students. If they will not be working on individual computers, be sure to explain any necessary modifications.

Distribute the activity to the students. Explain that in this activity they will use GIS to observe and analyze the locations and sizes of the world's ten largest cities in eight different time periods from 100 CE to the year 2005. They will identify changes in both location and size of the world's largest cities and speculate on possible reasons for the patterns they observe.

The following are things to look for while the students are working on this activity:

- Are the students using a variety of GIS tools?
- Are the students answering the questions?
- Are students asking thoughtful questions?

Concluding the lesson

Use the transparency from the CD to compare student ideas and observations from the activity. Summarize student observations on the transparency as they share and discuss their observations with the class. Use this discussion as a forum to elaborate on relevant themes in world history (such as the decline of the Roman Empire or the Industrial Revolution) and the value of using geography's spatial perspective to interpret the past. Ideally, this discussion should take place in the classroom with a projection device that displays the ArcMap map document as students discuss it. If this is not possible, you might want to conduct the discussion while students are still working on the computer.

Middle school assessment. Students will create a line graph of the most populous cities for the time periods studied and use the graph as a reference for writing an essay comparing two of the time periods. The essay will illustrate their understanding of the changes in spatial patterns of major population centers.

High school assessment. Students will create a line graph of the most populous cities for the time periods studied and use the graph as a reference for writing an essay comparing three or more time periods. The essay will illustrate their understanding of the changes in spatial patterns of major population centers. In addition, they will take the historical information from the map and their own research to make predictions about future locations of major population centers.

Extending the lesson

Challenge students to try the following:

- Explore the relationship between physical characteristics of the landscape and the locations of the world's most populated cities by adding layers reflecting world climate data (module 3) and ecoregion data (module 7).
- Conduct research on the historic cities and time periods mentioned in the lesson.
- Reflect on the questions this activity has raised in your mind and conduct further research and spatial analysis to answer those questions.
- Create ArcMap layouts of the historic time periods. Print these layouts and use them in reports about ancient European cities.
- Create hypothetical maps of the world's most populated cities in future years. Compare these predictions with the 2005 maps. Discuss population growth trends you foresee in the twenty-first century.

See the "Resources by Module" section of this book's Web site—www.esri.com/ourworldgiseducation—for print, media, and Internet resources on the topics of population, historical time periods, and ancient cities.

Answer key

Step 2: Look at cities in 100 CE

Q1.   Where are the ten largest cities in the world in 100 CE located on the earth's surface? Many of the cities are at about the same latitude (north of the Tropic of Cancer, or approximately 30 degrees north latitude). All of the cities are in the Northern and Eastern hemispheres.

Q2.   Where are they located in relation to each other? Five of the cities are located on the Mediterranean Sea. All but three of the cities are in Asia. None of the cities is located in North or South America or Australia.

Q3.   Where are they located in relation to physical features? All of the cities are located near rivers or near the coast.

Q4.   What are possible explanations for the patterns you see on this map? Answers will vary and may include the influence of climate, the extent of the Roman Empire, trade, suitability for agriculture.

Step 3: Find historic cities and identify modern cities and countries

Q5.   Use the Find and Identify tools to complete the information in the table below.

City		Modern country
Historic name	Modern name	
Carthage	Tunis	Tunisia
Antioch	Antioch	Turkey
Peshawar	Peshawar	Pakistan

Step 4: Find the largest city of 100 CE and label it

Q6.   What's your estimate of how many people lived in the world's largest city in 100 CE? Answers will vary.

Q7.   What was the largest city in 100 CE? Rome

Q8.   What was the population of the world's largest city in 100 CE? 450,000

Step 5: Look at cities in 1000 CE and label the most populous city

Q9.   What notable changes can you see from 100 CE to 1000 CE? All the cities have changes from 100 CE to 1000 CE. The Mediterranean Sea is no longer the site of half the world's largest cities.

Q10. What similarities can you see between 100 CE and 1000 CE?
      Cities still cluster around 30 degrees north latitude.
      The population of each of the cities is under 1,000,000.
      Nearly all of the cities are in Asia.
      None of the top 10 cities is located in the Americas or Australia.

Q11. What was the largest city in 1000 CE? Cordova

Q12. What was the population of the world's largest city in 1000 CE? 450,000

Step 6: Compare other historical periods and formulate a hypothesis

Q13. Complete the table below.

Year CE	Largest city	Population of largest city	Major differences in top 10 cities compared with previous time period
100	Rome	450,000	Not applicable.
1000	Cordova	450,000	Mediterranean Sea is no longer a center of urban development.
1500	Beijing	672,000	Four of the 10 largest cities are now in China.
1800	Beijing	1,100,000	A city exceeds 1,000,000 for the first time. Europe now has three of the largest cities. Japan now has three of the largest cities.
1900	London	6,480,000	Nine of the 10 largest cities are now on two continents Europe (six) and North America (three). Five of the European cities are in Western Europe. Only one of the cities is in Asia. All 10 cities are over 1,000,000. Major cities move into more northern latitudes. The size of the largest city is six times the size of the largest city 100 years earlier.
1950	New York	12,463,000	First time a South American and Southern Hemisphere city is on the top 10 list. Europe drops back to three of the top 10 cities.
2000	Tokyo	34,450,000	Half of cities are over 15,000,000. None of the top 10 cities is in Western Europe.
2005	Tokyo	35,197,000	Jakarta, Indonesia, has replaced Los Angeles, USA, on the top 10 list. The United States has only one of the top 10 cities.

Q14. Using the map document and your answers in Q13, identify historical periods associated with the greatest changes and provide possible explanations for the changes.

Time period of significant change	Explanation for change
1800 through 1900	The Industrial Revolution caused rapid growth of European and North American cities. It also led to the immigration of millions of people to the United States from Europe.
1950 through 2005	The size of the world's largest cities mushroomed because of rapid economic growth in the developed world and the loss of agricultural jobs in the developing world (sending people to cities in hopes of finding a job).

Step 7: Investigate cities in the present time

Q15. How many of your original guesses are among the Top 10 Cities, 2000 CE? Answers will vary.

Q16. Which cities did you successfully guess? Answers will vary.

Q17. In the table below, write the city's name, 2005 population, and rank. Possible answers are listed in the table below.

Q18. Continue to fill out the table for the other cities in your list or cities that interest you. If you have cities on your list that are not in the top 30, fill in the name, leave the Population column blank, and write >30 in the Rank column. Possible answers are listed in the table below.

City	2005 population	Rank
Los Angeles	12,298,000	12
Hong Kong		>30
Bombay	18,196,000	5
Mexico City	19,411,000	2
Chicago	8,814,000	24

Q19. In general, how far are these other cities from the top 10? Answers will vary.

Assessment rubrics

Middle school

	Standard	Exemplary	Mastery	Introductory	Does not meet requirements
1	The student knows and understands how to use maps to analyze spatial distributions and patterns	Creates an accurate and well-labeled graph of the most populous cities over time; uses GIS to analyze population patterns and how they change throughout time, compares this with his/her original predictions, and makes predictions on future trends	Creates an accurate graph of the most populous cities over time; uses GIS to analyze population patterns and how they change throughout time, and compares this with his/her original predictions	Creates a graph of the most populous cities over time that is inaccurate or mislabeled in some places; uses GIS to view location patterns of the major population centers	Does not complete a graph of the most populous cities over time; has difficulty identifying location patterns of major population centers
12	The student knows and understands the spatial patterns of settlement in different regions of the world	Describes in detail how major cities have changed and are influenced by their location relative to physical features, and elaborates on what these features are and whether they were a positive or negative force on the cities	Identifies how major cities have changed and are influenced by their location relative to physical features such as water and other major cities	Identifies how the major cities have spread throughout the various regions of the world	Identifies the locations of major cities but does not identify a pattern of settlement
17	The student knows and understands how the spatial organization of a society changes over time	Compares the major cities of at least two time periods and makes predictions as to why changes occurred over time; provides detailed evidence of how factors such as technology and transportation influenced life in these cities	Compares the major cities of two time periods and makes predictions as to why changes in these population centers occurred over time	Identifies characteristics of the major cities of two different time periods but does not draw comparisons between the two eras	Identifies a few characteristics of major cities of two different time periods or describes characteristics for only one time period

This is a four-point rubric based on the National Standards for Geographic Education. The mastery level meets the target objective for grades 5–8.

High school

	Standard	Exemplary	Mastery	Introductory	Does not meet requirements
1	The student knows and understands how to use geographic representations and tools to analyze and explain geographic problems	Creates an accurate and well-labeled graph of the most populous cities over time; uses GIS to analyze how population patterns change throughout time, compares this with his/her original thoughts, and makes predictions on future trends; uses GIS to create an original map to illustrate these points	Creates an accurate graph of the most populous cities over time; uses GIS to analyze how population patterns change throughout time, compares this with his/her original thoughts, and makes predictions on future trends	Creates a graph of the most populous cities over time that is inaccurate or mislabeled in some places; uses GIS to analyze population patterns and how they change throughout time	Does not complete a graph of the most populous cities over time; uses GIS to analyze population patterns, but has difficulty identifying how they change throughout time
12	The student knows and understands the functions, sizes, and spatial arrangements of urban areas	Elaborates, by using visual and textual materials, on characteristics of major cities throughout time and how factors such as trade routes and technology influenced the growth of these places	Identifies characteristics of major cities throughout time and how factors such as trade routes and technology influenced the growth of these places	Identifies characteristics of major cities throughout time	Identifies major cities and a few of their characteristics
17	The student knows and understands how processes of spatial change affect events and conditions	Cites specific examples of how changes in politics, transportation, etc., caused major population centers to shift from one location to another between three time periods	Identifies how changes in politics, transportation, etc., influenced the location of major population centers in three time periods	Lists one or two major cities that changed from one time period to the next and provides explanation for the change	Lists one or two major cities that changed from one time period to the next

This is a four-point rubric based on the National Standards for Geographic Education. The mastery level meets the target objective for grades 9–12.

91

Answer key to part A of the assessment

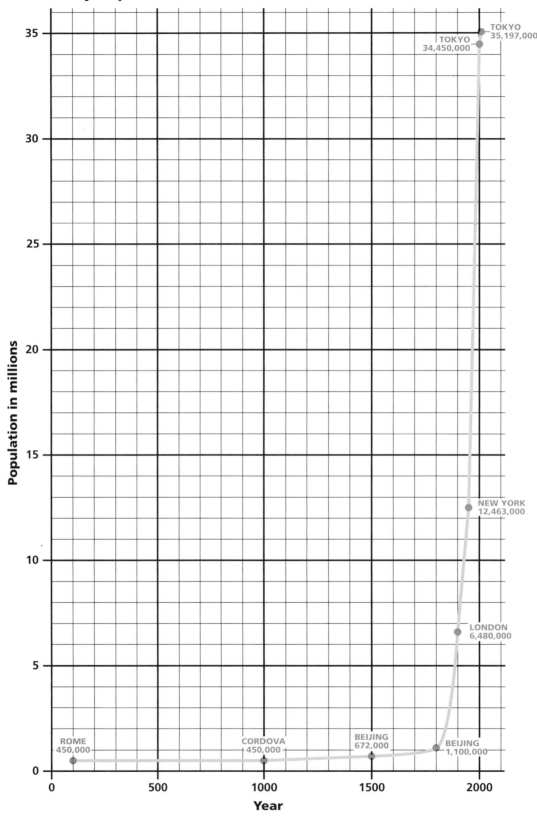

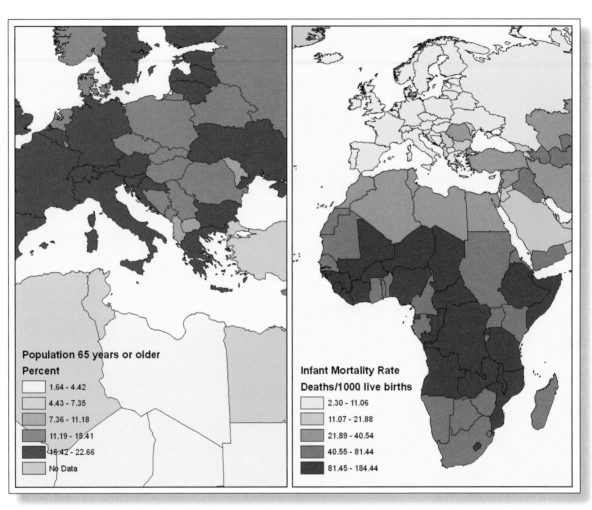

Maps of demographic data and standard-of-living indicators reveal differences between Europe and Africa.

Population 65 years or older
Percent
- 1.64 - 4.42
- 4.43 - 7.35
- 7.36 - 11.18
- 11.19 - 15.41
- 15.42 - 22.66
- No Data

Infant Mortality Rate
Deaths/1000 live births
- 2.30 - 11.06
- 11.07 - 21.88
- 21.89 - 40.54
- 40.55 - 81.44
- 81.45 - 184.44

Growing pains

A regional investigation of Europe and Africa

Lesson overview

In this lesson, students will compare the processes and implications of population growth in one of the world's fastest growing regions, sub-Saharan Africa, and the slowest growing region, Europe. Through the analysis of standard-of-living indicators in these two regions, students will explore some of the social and economic implications of rapid population growth.

Estimated time

Two to three 45-minute class periods

Materials

The student worksheet files can be found on the Data and Resources CD. Install the teacher resources folder on your computer to access them.

Location: OurWorld_teacher\Module4\Lesson2
• Student PDF: M4L2_student.pdf
• Student answer sheet: M4L2_student_answer_sheet.doc
• Student assessments: M4L2_assessment.pdf

Additional materials
• One sheet of drawing paper and one marker per student

Objectives

After completing this lesson, a student is able to do the following:

• Describe the fundamentals of population growth by explaining the relationship between birth rate, death rate, and natural increase
• Identify the fastest and slowest growing regions in the world today
• Explain the socioeconomic implications of rapid population growth
• Explain the slow population growth in Europe and how standard-of-living indicators are affected
• Explain the rapid population growth in areas of Africa and how standard-of-living indicators are affected

GIS tools and functions

 View attributes for a feature on a map

 Zoom in or out on the map

 Add a layer to the map

 Zoom to the full map extent

 Go back to the previous map extent

 View the entire layout page

 Change the layout template

 List layout templates

 Select, move, and resize layout elements

- Turn layers on and off
- Expand and activate a data frame
- Create and print two layouts using different data frames

National Geography Standards

Standard	Middle school	High school
1 How to use maps and other geographic representations, tools, and technologies to acquire, process, and report information from a spatial perspective	The student knows and understands how to use maps to analyze spatial distributions and patterns	The student knows and understands how to use geographic representations and tools to analyze and explain geographic problems
9 The characteristics, distribution, and migration of human populations on Earth's surface	The student knows and understands the demographic structure of a population	The student knows and understands trends in world population numbers and patterns
18 How to apply geography to interpret the present and plan for the future	The student knows and understands how to apply the geographic point of view to social problems by making geographically informed decisions	The student knows and understands how to use geographic knowledge, skills, and perspectives to analyze problems and make decisions

M ● ● ● ● ○ ○ ○
L ● ●

Teaching the lesson

Introducing the lesson

Begin the lesson with a discussion of world population growth. Remind students that the world's population reached six billion in 1999 and continues to grow.

Consider the following questions:

- Why is the earth's population growing?
- Are all regions growing at the same rate of speed?
- What does "overpopulation" mean, and at what point would you characterize the world as overpopulated?

If your community or state is growing, that would be a good starting point for this discussion. How has population growth affected your community or state? Do students see this as a good thing or a bad thing?

Student activity

We recommend that you complete the activity yourself before presenting the lesson in class. Doing so will allow you to modify the activity to accommodate the specific needs of your students. If they will not be working on individual computers, be sure to explain any necessary modifications.

Distribute the activity to students. Explain that in this activity they will use GIS to compare a region of the world where population is growing fast and a region where it is growing slowly. They will also investigate the relationship between a region's rate of population growth and its standard of living.

The following are things to look for while the students are working on this activity:

- Are the students using a variety of tools?
- Are the students answering the questions?
- Do students need help with the lesson's vocabulary?

Concluding the lesson

When the class has finished the activity, give each student a piece of drawing paper on which to write his or her hypothesis from Q17. Students should tape their hypotheses to the wall or blackboard and discuss them. Look for similarities and differences among the hypotheses. Allow students to question each other to clarify confusing or contradictory statements. If possible, try to reach consensus about the relationship between a country's rate of natural increase and its standard of living based on evidence from the activity.

Middle school assessment. Students will play the role of a special liaison to the United Nations, in charge of establishing a partnership between a slow growing nation and a fast growing nation. They will need to identify issues critical to each country and devise a way the countries can form a partnership to improve their standards of living.

High school assessment. Students will play the role of a special liaison to the United Nations, in charge of establishing a partnership between a group of slow growing nations and a group of fast growing nations. They will need to identify issues critical to each group and devise a way the countries can form a partnership to improve their standards of living.

Extending the lesson

Challenge students to try the following:

- Explore and map additional attributes from the module 4 data folder. Look for additional social and economic implications of rapid population growth.
- Test your hypothesis about the relationship between standard of living, net migration, and the rate of natural increase by investigating countries in other parts of the world.
- Explore an African country's population, birth rate, death rate, and standard-of-living indicators and write a report about the country, including a map layout.
- Use ArcMap attribute queries to identify countries that do not match your hypothesis. Try to figure out an explanation for these anomalies.
- Explore gender differences in standard of living in Europe and/or sub-Saharan Africa. Using the module data, map and analyze male and female life expectancies, male and female literacy rates, and male and female infant and child mortalities.
- Conduct research on the impact of HIV/AIDS on death rates and population growth in sub-Saharan Africa.

See the "Resources by Module" section of this book's Web site—www.esri.com/ourworldgiseducation—for print, media, and Internet resources on the topics of Africa, Europe, demographics, and standard-of-living indicators.

Answer key

Step 2: Compare birth rate and death rate data

Q1. Which world region or regions have the highest birth rates? Africa, South Asia

Q2. Which world region or regions have the lowest birth rates? North America, Europe, Russia, Australia

Q3. Which world region or regions have the highest death rates? Africa

Q4. Which world region or regions have the lowest death rates? Mexico, Central America, western South America, Northern Africa, Southwest Asia

Q5. If the overall rate of growth is based on the formula BR – DR = NI, which world regions do you think are growing the fastest? Areas of sub-Saharan Africa, Arabia, and Central America

Q6. Which world regions do you think are growing the slowest? Many European countries, Russia

Q7. Choose two European countries and two African countries and record their birth and death rates in the table below. Answers will vary and may include the following:

Country and continent	Birth rate/1,000	Death rate/1,000
Niger (Africa)	50.16	20.59
Spain (Europe)	9.98	9.81
Hungary (Europe)	9.66	13.05
Ethiopia (Africa)	37.39	14.67
Chad (Africa)	45.30	16.04

Q8. List three questions that the Birth Rate and Death Rate maps raise in your mind. Answers will vary.

Step 3: Add the Natural Increase layer

Q9. What is happening to the populations of countries that are pink? Their death rates exceed their birth rates. Over time these populations will decline unless migration into the countries makes up for the negative natural increase.

Q10. Which world regions are growing the fastest? Sub-Saharan Africa and Southwest Asia

Q11. Which world regions are losing people or not growing? Eastern Europe, Russia, and southern Africa

Q12. Think about what it would mean for a country to have a population that is growing rapidly or one that is growing slowly or shrinking. Which of these two situations do you think would cause more problems within the country? Students may answer that either situation or both situations can lead to problems.

List some of the problems you would expect to see. Answers will vary, but students who answered that rapid growth would cause more problems should recognize that a country with rapid growth will have a difficult time keeping up with the constantly increasing need for education, health care, social infrastructure, resources, and jobs. Students who answered that a shrinking population would cause more problems should list effects such as inability to defend itself, loss of industry, inability to fill industrial or technology jobs, the pressure to increase immigration, and so on.

Step 4: Look at standard-of-living indicators for Europe and Africa

Q13. Complete the table below.

Indicator	Compare sub-Saharan Africa and Europe	What does this indicate about the standards of living in these regions?
Population 65 years or older	Sub-Saharan Africa: Most countries have 1.64% to 4.42% in this age group. Europe: Most countries have 15.42% to 22.66% in this age group.	Africa: Low percent indicates many people die prematurely and do not reach old age. Low standard of living. Europe: High percent indicates more people live into their sixties and beyond because of good health care, sanitation, adequate food supply, and so forth. High standard of living.
GDP per capita	Sub-Saharan Africa: Most countries have the lowest level of GDP. Europe: Most countries have the second-highest or highest level of GDP.	This is not the same as average income—be sure that students do not make that assumption. A higher GDP per capita does indicate a wealthier country, and that means more money to spend on the infrastructure. High GDP means a high standard of living and enough capital to continue to grow and expand economically.
Infant mortality rate	Sub-Saharan Africa: All countries have 21.89 or more infant deaths/1,000 born. Europe: Almost all countries have fewer than 21.89 infant deaths/1,000 born.	High rate of infant mortality indicates a low standard of living. This statistic is typically used to evaluate the health conditions (sanitation, health care, food supply, disease, etc.) in a country because newborns are much more susceptible to death from health hazards than adults or older children.
Life expectancy	Sub-Saharan Africa: Most countries have a life expectancy of less than 57.87 years. Europe: Most countries have the highest life expectancy, 75.35–83.52 years.	A higher life expectancy indicates a higher standard of living because it reflects the presence of conditions that sustain life and/or a relatively low presence of threats to life.
Literacy rate	Sub-Saharan Africa: Only a few countries have literacy rates above 84%. Europe: Almost all countries have literacy rates of at least 94.11%.	Information on literacy, while not a perfect measure of education in a country, is probably the most easily available and valid for international comparisons. Low levels of literacy and education in general can impede the economic development of a country in the current rapidly changing, technology-driven world.
Percent of workforce in service sector	Sub-Saharan Africa: The service sector of the workforce is below 61% in most countries. Europe: The service sector of the workforce is above 61% in most countries.	A higher percent of the workforce in the service sector indicates a higher standard of living. As a country becomes more developed economically, a larger percent of its workforce is employed in the service sector. The workforce in less developed countries is characterized by higher percent of workers in agriculture and industry.

Step 5: Add the Net Migration layer

Q14. In Q13 you compared standard-of-living indicators in Europe and sub-Saharan Africa. Based on your observations of those indicators, which region would you expect to have a negative net migration? A positive net migration?
Negative: Sub-Saharan Africa (countries with low standards of living)
Positive: Europe (countries with high standards of living)
Explain your answers: Answers will vary, but students should recognize that there will be more out-migration from countries with low standards of living and more in-migration to countries with a higher standard of living.

Q15. Summarize the overall patterns of net migration in Europe and sub-Saharan Africa in the table below.

Net migration in sub-Saharan Africa	Net migration in Europe
There is no clear pattern. Some countries have net in-migration, and others have net out-migration.	Western Europe has net in-migration, and most Eastern European countries have net out-migration.

Q16. What political or social conditions or events could explain any of the migration patterns you see on the map? Possible answers include Balkan wars, fall of communism, higher standard of living in Western Europe, and the end of civil war in Liberia.

Step 6: Draw conclusions

Q17. Based on your map investigations, write a hypothesis about how a country's rate of natural increase affects its standard of living and its net rate of migration. Answers will vary, but students should note that natural increase has a direct effect on standard of living and that standard of living creates push–pull factors that influence migration.

Q18. In the table below, illustrate your hypothesis with data from one European country and one sub-Saharan African country. Answers will vary depending on the hypothesis that was formulated in Q17. However, students should include natural increase, net migration, and other data that support their hypotheses.

Europe	Data	Africa
	Country name	
	Natural increase	
	Net migration	

Step 7: Design a layout

Q19. What are the units of measurement? Kilometers

Assessment rubrics

Middle school

	Standard	Exemplary	Mastery	Introductory	Does not meet requirements
1	The student knows and understands how to use maps to analyze spatial distributions and patterns	Uses GIS to compare and analyze growth and demographic trends in two countries; makes predictions from the data on future population trends; creates original maps to support findings	Uses GIS to compare and analyze growth and demographic trends in two countries; creates views that isolate regions of slow and fast growth; creates maps to support findings	Uses GIS to identify regions and countries with slow and fast growth rates; printed maps do not support findings	Has difficulty identifying demographic patterns; does not print any maps
9	The student knows and understands the demographic structure of a population	Analyzes demographic data of two countries with fast and slow growth rates and draws conclusions on how the population structure of each country came to exist	Identifies and analyzes data that illustrates the demographic makeup of two countries with slow and fast growth rates	Identifies demographic characteristics of countries with slow and fast growth rates	Has difficulty identifying differences between countries with slow and fast growth rates
18	The student knows and understands how to apply the geographic point of view to social problems by making geographically informed decisions	Identifies critical growth issues for a country of fast growth and a country of slow growth; creates a detailed program through which the countries could establish a partnership to solve growth-related issues; elaborates on how this program could be replicated by other countries	Identifies critical growth issues for a country of fast growth and a country of slow growth; determines several ways that the countries could form a partnership for the mutual benefit of both in regard to growth issues	Identifies critical growth issues for a country of fast growth and another of slow growth; determines one or two ways the countries could form a partnership	Identifies critical growth issues for two countries but does not address the issue of establishing a partnership between them

This is a four-point rubric based on the National Standards for Geographic Education. The mastery level meets the target objective for grades 5–8.

High school

Standard	Exemplary	Mastery	Introductory	Does not meet requirements
1 The student knows and understands how to use geographic representations and tools to analyze and explain geographic problems	Uses GIS to compare and analyze growth and demographic trends in countries throughout the world; makes predictions from the data provided and additional sources on future population trends; creates original maps to support findings	Uses GIS to compare and analyze growth and demographic trends in countries throughout the world; makes predictions from the data on future population trends; creates maps to support findings	Uses GIS to compare and analyze growth and demographic trends in countries throughout the world; makes predictions from the data on future population trends; printed maps do not support findings	Attempts to make comparisons between countries using demographic data in a GIS; does not print any maps
9 The student knows and understands trends in world population numbers and patterns	Creates two groups of countries, each with similar demographic trends, including standard-of-living indicators; one group represents fast-growth countries, and the other represents slow-growth countries; makes predictions on how these trends will change through time	Creates two groups of countries, each with similar demographic trends, including standard-of-living indicators one group represents fast-growth the other represents slow-growth countries	Creates two groups of countries with similar growth characteristics; one group represents fast-growth countries, and the other represents slow-growth countries	Identifies one or two countries that represent either slow or fast growth
18 The student knows and understands how to use geographic knowledge, skills, and perspectives to analyze problems and make decisions	Creates a report that establishes a coalition of slow- and fast-growth countries that work together for the mutual benefit of all involved; the report takes into account issues critical to the countries that participate and elaborates on possible solutions	Creates a report that establishes a coalition of slow- and fast-growth countries that work together for the mutual benefit of all involved; the report takes into account issues critical to the countries that participate	Lists several issues that are critical to fast- and slow-growth countries; attempts to find ways in which the countries can partner	Identifies one or two issues critical to countries with slow growth and those with fast growth

This is a four-point rubric based on the National Standards for Geographic Education. The mastery level meets the target objective for grades 9–12.

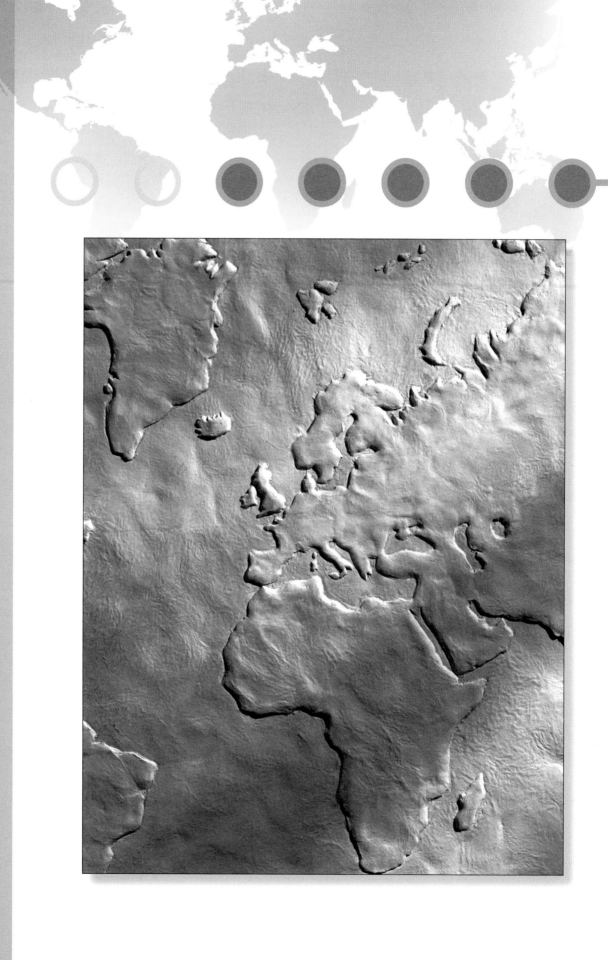

MODULE 5

Boundaries

Lesson 1: Crossing the line: A global perspective

Students will explore the nature and significance of international political boundaries. Through an investigation of contemporary political boundaries, they will identify boundary types, compare patterns of territorial morphology (size and shape), and explore the relationship of boundaries to national cohesiveness and economic potential. By comparing 1992 and 2007 world political boundaries, students will observe the evolution of boundaries over time.

Lesson 2: A line in the sand: A regional investigation of Saudi Arabia and Yemen

Students will study the creation of a new border between Saudi Arabia and Yemen. Using data from the June 2000 Treaty of Jeddah, they will draw the new boundary established by the treaty and analyze the underlying physiographic and cultural forces that influenced the location of that boundary. In the process they will come to understand how any map of the world must be considered a tentative one, as nations struggle and cooperate with each other.

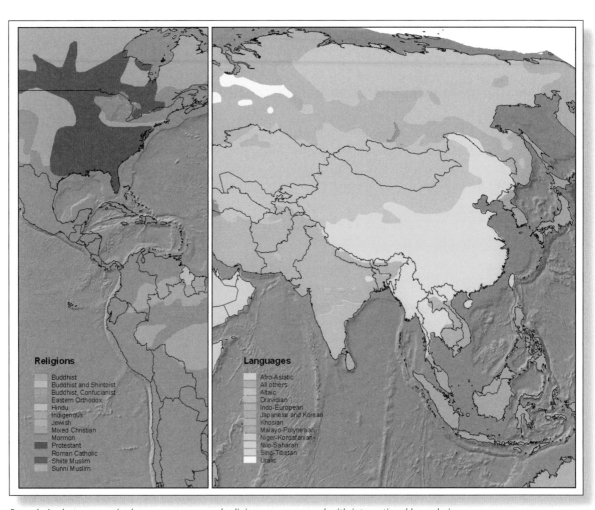

Religions

- Buddhist
- Buddhist and Shintoist
- Buddhist, Confucianist
- Eastern Orthodox
- Hindu
- Indigenous
- Jewish
- Mixed Christian
- Mormon
- Protestant
- Roman Catholic
- Shiite Muslim
- Sunni Muslim

Languages

- Afro-Asiatic
- All others
- Altaic
- Dravidian
- Indo-European
- Japanese and Korean
- Khosian
- Malayo-Polynesian
- Niger-Kordafanian
- Nilo-Saharan
- Sino-Tibetan
- Uralic

Boundaries between major language groups and religions are compared with international boundaries.

Module 5: Lesson 1

Crossing the line
A global perspective

Lesson overview

Students will explore the nature and significance of international political boundaries. Through an investigation of contemporary political boundaries, they will identify boundary types, compare patterns of territorial morphology (size and shape), and explore the relationship of boundaries to national cohesiveness and economic potential. By comparing 1992 and 2007 world political boundaries, students will observe the evolution of boundaries over time.

Estimated time

Two 45-minute class periods

Materials

The student worksheet files can be found on the Data and Resources CD. Install the teacher resources folder on your computer to access them.

Location: OurWorld_teacher\Module5\Lesson1
- Student PDF: M5L1_student.pdf
- Student answer sheet: M5L1_student_answer_sheet.doc
- Student assessments: M5L1_assessment.pdf

Additional materials
- A world atlas or map of Europe showing names of mountain ranges

Objectives

After completing this lesson, a student is able to do the following:

- Define and give examples of physiographic, geometric, and anthropographic boundaries
- Describe the political and economic implications of a country's size and shape
- Explain the relationship between boundary characteristics and national cohesiveness
- Explain international boundary changes that occurred in the late twentieth century

GIS tools and functions

Zoom to a specific area on the map

Draw a line

Change the line color

Identify a feature on the map

Select a graphic

Move the map to bring a different portion of it into view

Zoom to full extent

Add layers to the map

Find a feature in a layer, select it, and zoom to it

Save the map document

- Turn layers on and off
- Expand and collapse legends in the table of contents

National Geography Standards

Standard	Middle school	High school
3 How to analyze the spatial organization of people, places, and environments on Earth's surface	The student knows and understands how to use the elements of space to describe spatial patterns	The student knows and understands how to apply concepts and models of spatial organization to make decisions
13 How the forces of cooperation and conflict among people influence the division and control of Earth's surface	The student understands how cooperation and conflict among people contribute to economic and social divisions of the Earth's surface	The student knows and understands why and how cooperation and conflict are involved in shaping the distribution of social, political, and economic spaces on Earth at different scales
18 How to apply geography to interpret the present and plan for the future	The student knows and understands how varying points of view on geographic context influence plans for change	The student knows and understands contemporary issues in the context of spatial and environmental perspectives

M ● ● ● ● ● ○ ○
L ● ○

Teaching the lesson

Introducing the lesson

Write the following quotation on the board or on a transparency:

> "When you go around the Earth in an hour and a half . . . you look down there and you can't
> imagine how many borders and boundaries you cross, again and again and again, and you
> don't even see them . . . from where you see it, the thing is a whole, and it's so beautiful."
> Russell L. Schweickart
> Apollo 9, March 3–13, 1969
> (From *The Overview Effect*, 1998)

Use this quotation as a springboard to discuss the following questions:

- What are boundaries?
- Who draws the boundary lines?
- What purpose do boundaries serve?
- If boundaries are invisible lines, how do you know when you've crossed one?
- Once you've crossed one of these invisible lines, what has changed?
- Can you think of any problems that boundaries may cause?

Throughout the discussion, emphasize that although most boundaries are unmarked and invisible, they determine our perception of spaces and places on the earth. Boundaries between countries help maintain order in the world because they define internationally recognized and sovereign political entities. Conflict can result when boundary lines are disputed. Boundaries are also potential sources of conflict because they are the point of contact between neighboring people.

Challenge students to identify places in the world where international boundaries have changed or are in conflict. What do students know about the reasons for those boundary changes and conflicts? Tell the class that they are going to do a GIS investigation that will explore the characteristics of modern international boundaries and investigate recent boundary changes.

Student activity

We recommend that you complete the activity yourself before presenting the lesson in class. Doing so will allow you to modify the activity to accommodate the specific needs of your students. If they will not be working on individual computers, be sure to explain any necessary modifications.

Distribute the activity to the students. Explain that in this activity, they will use GIS to investigate different types of international boundaries, explore the implications of various boundary configurations, and observe recent boundary changes.

The following are things to look for while the students are working on this activity:

- Are the students using a variety of GIS tools?
- Are the students answering the questions?
- Are students asking thoughtful questions?

This activity contains instructions for students to periodically stop and ask the teacher how to save their work. These are good spots to stop the class for the day and to pick up the investigation the next day. You may want to inform your students in advance how they should rename their map document and where to save it.

Concluding the lesson

Use a projection device to display the **Global5** map document in the classroom. As a group, compare student observations and conclusions from the lesson. Students can take turns being the "driver" on the computer to highlight boundaries and observations that are identified by members of the class. Focus on the following aspects of the activity in your discussion:

- Where did students observe the coincidence of political and physiographic boundaries?
- What examples of territorial morphology (country shapes and sizes) did the students find? Ask students to speculate on ways that a country's size and shape could influence its cohesiveness.
- How can a country's boundaries influence its economic advantage?
- What kinds of problems are likely to arise when political and anthropographic (cultural) boundaries do not coincide?
- What is the nature of the boundary changes that occurred between 1992 and 2007? Based on student responses in Q21, which of the new countries do students believe are in the strongest position today in terms of cohesiveness?

Middle school assessment. Students have to identify a current international boundary that they predict could change in the next 25 years, prepare a map of the projected boundary change, and describe the impact of the change.

High school assessment. Students have to identify two current international boundaries that they predict could change in the next 25 years (one involving the splitting of a country and one involving a merging of countries), prepare a map of the projected boundary changes, and compare the effects of the changes.

Extending the lesson

Challenge students to try the following:

- Conduct research on world boundary changes during the twentieth century. Use GIS software to prepare a sequence of layouts reflecting those changes.
- Explore the nature of political boundaries in your own community and state. What kinds of boundaries are they? How does the shape of your town or state affect its cohesiveness? What are the economic advantages and disadvantages of your town's or state's boundary configuration? How have your town or state's boundaries changed over time?
- Search newspapers and magazines (both on- and offline) for coverage of border conflicts and related issues around the world. Create a GIS map to illustrate these conflicts.

See the "Resources by Module" section of this book's Web site—www.esri.com/ourworldgiseducation— for print, media, and Internet resources on the topic of political boundaries.

Answer key

Step 2: Explore mountain ranges as physiographic boundaries

Q1.   The Pyrenees Mountains are the border between which two countries? Spain and France

Q2.   Complete the table below. Possible answers are listed in the table.

Countries that have mountain ranges as political boundaries	Mountains that form the boundary
Italy and Switzerland	Alps
Italy and France	Alps
Poland and Czech Republic	Sudeten Mountains

Step 3: Explore bodies of water as physiographic boundaries

Q3.   Record the names of three pairs of countries that share a boundary that's a river. Possible answers are listed in the table.

Countries that have a river as a boundary	River that forms the boundary
France and Germany	Rhine
Romania and Bulgaria	Danube
Romania and Serbia	Danube
Belarus and Ukraine	Dnieper
Switzerland and Germany	Rhine

Q4.   Name three landlocked countries in Western Europe. Possible answers: Switzerland, Austria, Luxembourg, Czech Republic, Slovakia, Hungary, Serbia.

Step 4: Explore geometric boundaries

Q5.   List three pairs of countries with a shared geometric boundary. Possible answers:
Egypt and Sudan
Sudan and Chad
Libya and Chad
Niger and Algeria
Libya and Sudan
Algeria and Mali
Namibia and Botswana
Algeria and Mauritania
Angola and Zambia

Q6.   Write the new name you gave the map document and where you saved it.
Document example: ABC_Global5.mxd
Location example: C:\Student\ABC

Step 5: Explore anthropographic boundaries based on language and religion

Q7.   Determine the principal language group in South America and Western Europe.
South America: Indo-European
Western Europe: Indo-European

Q8. Use the Identify and Zoom tools to locate countries separated by an anthropographic boundary based on language. List three pairs of such countries. Possible answers:
India and China
Georgia and Russia
North Korea and China
Kazakhstan and Russia
Finland and Sweden
Brazil and Paraguay
Thailand and Laos
Botswana and Zimbabwe

Q9. Determine the principal religions in North America and Africa.
North America: Protestant, Roman Catholic, Indigenous
Africa: Sunni Muslim, Indigenous

Q10. Use the Identify and Zoom tools to locate countries separated by an anthropographic boundary based on religion. List three pairs of such countries. Possible answers:
India and China/Myanmar/Pakistan/Bangladesh
Thailand and Malaysia
Vietnam and Cambodia/Laos
Kazakhstan and Russia
Iran and Pakistan/Iraq/Turkmenistan
Finland and Russia
Ireland and United Kingdom
Germany and Czech Republic

Step 6: Review physiographic, geometric, and anthropographic boundaries

Q11. List additional examples of countries separated by physiographic, geometric, or anthropographic boundaries for each continent in the table. See Q5, Q8, and Q10 for possible answers.

Step 7: Explore the effects of boundary shape, cultural diversity, and access to natural resources

Q12. Locate another example of each type of country. Record the countries in the Example 2 column. Possible answers are listed in the table.

Type of country	Example 1	Example 2
Elongated	Chile	Vietnam, Panama
Fragmented	Philippines	Indonesia, Japan
Circular/hexagonal	France	Uruguay, Zimbabwe
Small/compact	Bulgaria	Costa Rica, Belgium
Perforated (has a "doughnut hole")	South Africa	Italy
Prorupted (has a "panhandle")	Namibia	Thailand, Afghanistan

Q13. Identify three culturally uniform countries on the basis of language group. Possible answers: Japan, France, Argentina, Italy, Hungary.

M ● ● ● ● ● ○ ○
L ● ○

Q14. Identify three culturally diverse countries on the basis of language groups. Possible answers: Canada, Spain, Nigeria, Burkina Faso, Syria, Turkey, India, Sudan, Sri Lanka, Namibia.

Q15. Record an example of a landlocked country for each of the following continents. For a continent that does not have a landlocked country, write "None."

Continent	Landlocked country
North and Central America	None
South America	Possible answers: Bolivia, Paraguay
Africa	Possible answers: Mali, Burkina Faso, Niger, Chad, Central African Republic, Rwanda, Burundi, Uganda, Zambia, Zimbabwe, Botswana, Lesotho, Ethiopia, Malawi, Swaziland
Asia	Possible answers: Afghanistan, Nepal, Bhutan, Laos, Mongolia, Kyrgyzstan, Tajikistan, Armenia

Q16. Name two Southeast Asian countries that do not have any oil and gas resources within their land borders. Possible answers: Cambodia, Laos, Vietnam

Q17. Name two Southeast Asian countries that have oil and gas resources within their land borders. Indonesia, Brunei, Thailand, Malaysia, Myanmar

Step 8: Explore boundary changes that occured in the 1990s

Q18. Describe three political-boundary changes between 1992 and 2007.
Answers will vary but should focus on the changes in Eastern Europe and the former Union of Soviet Socialist Republics (USSR).

Q19. Name two countries that existed in 1992 but do not exist in 2007. Possible answers: USSR, Czechoslovakia, Yugoslavia

Q20. Write the new name you gave the map document and where you saved it.
Document Example: ABC_Global5.mxd
Location Example: C:\Student\ABC

Q21. Select three countries from group A and three from group B and complete the table.
Answers will vary. One possible answer from each group is given below.

Group	Country	Type of boundaries	Shape	Economic advantages or disadvantages	Likelihood of cohesiveness or of splitting apart
A	Czech Republic	Physiographic	Compact	Has more neighbors to trade with but fewer workers and natural resources.	Unlikely to split apart. (More cohesive than before, when it was part of elongated Czechoslovakia.)
B	Uzbekistan	Physiographic, geometric	Elongated, prorupted	Transportation challenges due to mountains or need to cross into another country. Desert separates two "green" areas of the country.	Has the potential to split apart. The panhandle region may combine with Kyrgyzstan.

113

Assessment rubrics

Middle school

	Standard	Exemplary	Mastery	Introductory	Does not meet requirements
3	The student knows and understands how to use the elements of space to describe spatial patterns	Using a variety of data, creates a GIS map to illustrate a predicted international boundary change	Using a variety of data, creates a map to illustrate a predicted international boundary change	Creates a map to illustrate a predicted international boundary change; provides some data to support ideas	Describes a change in a border but does not provide a map and has little or no data to support ideas
13	The student understands how cooperation and conflict among people contribute to economic and social divisions of Earth's surface	Using maps, describes how the proposed boundary change will affect the countries involved, including their roles in the global economy; provides sufficient data to support ideas	Describes how the proposed boundary change will affect the countries involved, including their roles in the global economy; provides sufficient data to support ideas	Attempts to describe how the proposed boundary change will affect the countries involved but does not address effects on the global economy; provides some data to support ideas	Does not address economic issues for the countries involved in the boundary change
18	The student knows and understands how varying points of view on geographic context influence plans for change	Describes the perspectives of the countries involved in the boundary change; addresses a variety of political and cultural issues and effects on the global community	Describes the perspectives of the countries involved in the boundary change; addresses a variety of political and cultural issues	Attempts to describe the perspectives of the countries involved in the boundary change; addresses only one issue	Describes only one country's perspective on the proposed boundary change

This is a four-point rubric based on the National Standards for Geographic Education. The mastery level meets the target objective for grades 5–8.

High school

	Standard	Exemplary	Mastery	Introductory	Does not meet requirements
3	The student knows and understands how to apply concepts and models of spatial organization to make decisions	Using a variety of data, creates a GIS map illustrating the merging of countries together and the splitting of countries apart; writes a thorough analysis of the projected changes and their impact using data to support ideas	Creates a map illustrating the merging of countries together and the splitting of countries apart; writes an essay describing the projected changes and supports ideas with data	Creates a map illustrating one type of boundary change; speculates as to some possible reasons and impacts	Describes a border change but does not provide a map and has little or no data to support ideas
13	The student knows and understands why and how cooperation and conflict are involved in shaping the distribution of social, political, and economic spaces on Earth at different scales	Using maps and writing, describes how the proposed boundary changes will affect the countries involved, including their roles in the global economy; provides sufficient data to support ideas	Describes how the proposed boundary changes will affect the countries involved and their roles in the global economy; provides sufficient data to support ideas	Attempts to describe how the proposed boundary changes will affect the countries involved but does not address effects on the global economy; provides some data to support ideas	Does not address economic issues for the countries involved in the boundary changes
18	The student knows and understands contemporary issues in the context of spatial and environmental perspectives	Describes the perspectives of the countries involved in the proposed boundary changes; addresses a variety of political and cultural issues and effects on the global community	Describes the perspectives of the countries involved in the proposed boundary changes; addresses a variety of political and cultural issues	Attempts to describe the perspectives of the countries involved in the proposed boundary changes; addresses only one issue	Describes only one country's perspective on a proposed boundary change

This is a four-point rubric based on the National Standards for Geographic Education. The mastery level meets the target objective for grades 9–12.

115

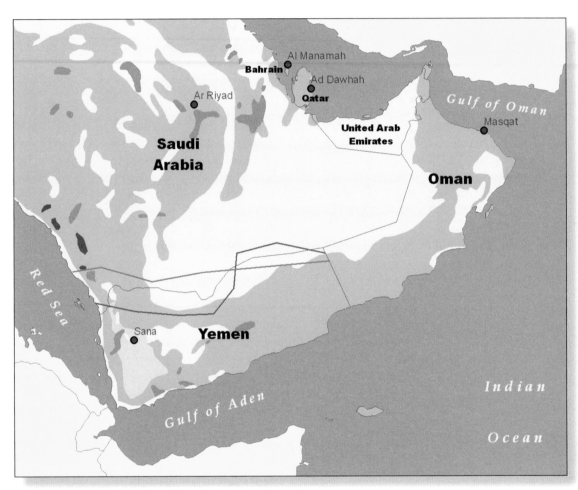

The border between Yemen and Saudi Arabia has changed over time.

A line in the sand

A regional investigation of Saudi Arabia and Yemen

Lesson overview

Students will study the creation of a new border between Saudi Arabia and Yemen. Using data from the June 2000 Treaty of Jeddah, they will draw the new boundary established by the treaty and analyze the underlying physiographic and cultural forces that influenced the location of that boundary. In the process they will come to understand how any map of the world must be considered a tentative one as nations struggle and cooperate with each other.

Estimated time

Three to four 45-minute class periods

Materials

The student worksheet files can be found on the Data and Resources CD. Install the teacher resources folder on your computer to access them.

Location: OurWorld_teacher\Module5\Lesson2
* Student PDF: M5L2_student.pdf
* Student answer sheet: M5L2_student_answer_sheet.doc
* Student assessments: M5L2_assessment.pdf

Objectives

After completing this lesson, a student is able to do the following:

* Describe the physical and population characteristics of the Arabian Peninsula
* Define and describe the Empty Quarter
* Explain major elements of the Treaty of Jeddah boundary agreement between Saudi Arabia and Yemen
* Identify physical and cultural characteristics of the Arabian Peninsula that are reflected in the new Saudi–Yemeni border agreement

GIS tools and functions

⊕ Zoom in on the map

⬅ Zoom to the previous extent

➕ Add layers to the map

⛶ Zoom in or out on the center of the map

〰 Draw a line

▶ Select a graphic

📂 Specify a location to save work

✎ Display and work with the Edit toolbar

✏ Create a line feature by adding points at specific latitude/longitude coordinates

↶ Undo an editing mistake

💾 Save the map document

▦ ◩ Select a map feature from a specific layer; clear the feature selection

📦 Show and hide the ArcToolbox window

- Use a bookmark
- Use MapTips to identify features
- Turn layers on and off and expand and collapse their legends
- Activate a data frame
- Export a layer to a geodatabase
- Set selectable layers
- Create a buffer
- Print a map

National Geography Standards

Standard	Middle school	High school
1 How to use maps and other geographic representations, tools, and technologies to acquire, process, and report information from a spatial perspective	The student understands how to use maps to analyze spatial distributions and patterns	The student understands how to use geographic representations and tools to analyze and explain geographic problems
4 The physical and human characteristics of places	The student understands how physical processes shape places and how different human groups alter the characteristics of places	The student understands the changing human and physical characteristics of places
13 How the forces of cooperation and conflict among people influence the division and control of Earth's surface	The student understands the multiple territorial divisions of the student's own world	The student understands why and how cooperation and conflict are involved in shaping the distribution of social, political, and economic spaces on Earth at different scales
18 How to apply geography to interpret the present and plan for the future	The student understands how various points of view on geographic context influence plans for change	The student understands contemporary issues in the context of spatial and environmental perspectives

M ● ● ● ● ● ○ ○

L ● ●

Teaching the lesson

Introducing the lesson

Divide the class into small groups. Explain to students that the lesson is about drawing boundary lines. In order to identify some of the important considerations in the demarcation of boundaries, each group will take five minutes to consider the following hypothetical scenario.

Size limitations in the school building require that your classroom be divided to create two new, smaller classrooms. Other than the wall dividing the two classrooms, there will be no new construction. Each group is charged with two tasks:

1. Identify the features of the present classroom that are valuable to the teachers and students who use that room (windows, for example).
2. Suggest a possible boundary line to divide the classroom and identify the features of the original classroom that each of the new classrooms will have.

When five minutes have passed, make a list on the blackboard or an overhead projector of the valuable classroom features that students identified in the first step. Let each group report on the boundary they propose. Use this activity as a springboard for a discussion of the issues involved in the creation of national boundaries. Be sure to include the following points in the discussion:

- Certain features of the physical environment have greater value than others to the people who will occupy and use that space.
- The human uses of a place influence the perceived value of its physical features.
- When a boundary line is drawn, it may not be possible to divide the valuable features evenly between the parties involved.

Tell the class that this activity will explore a twenty-first century case of the demarcation of a boundary between two countries. Although at a much different scale, this decision involved some of the same issues they faced in drawing a hypothetical boundary in their classroom.

Student activity

We recommend that you complete the activity yourself before presenting the lesson in class. Doing so will allow you to modify the activity to accommodate the specific needs of your students. If they will not be working on individual computers be sure to explain any necessary modifications.

Explain that in this activity they will use GIS to explore a region of the world where a boundary dispute has been recently settled after 65 years of conflict. They will explore alternatives for boundaries between the countries involved and analyze the underlying physiographic and cultural considerations that played a part in the resolution of that conflict.

The following are things to look for while the students are working on this activity:

- Are the students using a variety of tools?
- Are they answering the questions?
- Are they experiencing any difficulty managing the display of information in their map as they turn layers on and off?
- Are students experiencing any difficulty plotting latitude/longitude points or finishing their sketch when creating the Saudi–Yemeni boundary line feature?

Decide ahead of time where you want students to save the data for the new boundary line they will create. Students can export a feature class to the MiddleEast geodatabase if they have their own copy of the module 5 folder. Otherwise, you may want students to export a shapefile to another location.

Concluding the lesson

Refer your students to the activity that introduced this lesson: the creation of a hypothetical boundary line that divides their classroom into two new rooms. Review their conclusions and ask them to identify parallel issues in the settlement of the boundary dispute between Saudi Arabia and Yemen.

- Certain features of the physical environment have greater value than others to the people who will use that space. On the Arabian Peninsula, areas that get enough precipitation for agriculture, areas of grassland for grazing, sources of water, and areas with the strategic advantage of mountain peaks have greater value.
- The human uses of a place influence the perceived value of its physical features. On the Arabian Peninsula, livestock herding and farming are examples of traditional human uses.
- When a boundary line is drawn, it may not be possible to divide the valuable features evenly between the parties involved. On the Arabian Peninsula, most of the areas that get enough precipitation for agriculture, areas of grassland for grazing, and sources of water went to Yemen.

Ask students to identify issues that played a role in the Saudi–Yemeni border conflict that were not present in the classroom boundary scenario. For example:

- Historic boundaries and patterns of political control in the region played an important part in the Saudi–Yemeni border conflict. Discuss important historical events and periods such as the Ottoman Empire, the consequences of World War I in this region, international interest in the region during the twentieth century, and the British Protectorate of Aden.
- Nomadism and strong identity with regional tribal traditions are at odds with the delineation of a fixed boundary in this region. Discuss the various factors that influence a community's or a region's sense of itself.

Students may wonder why the Saudis were willing to yield so much territory to Yemen. Ask them to speculate on possible reasons for this apparent generosity. The Treaty of Jeddah states that the two countries will negotiate if sources of "shared natural wealth" are discovered in the border region. This means that Saudi Arabia reserves the right to reopen negotiations in the event that something that they value very highly—oil or gas, for example—is discovered near the new boundary. Also, Saudi Arabia has long been interested in constructing a pipeline to the Arabian Sea across the southern part of the peninsula. The Saudis may have been hoping that a generous settlement with Yemen on the border issue could make the Yemenis more willing to agree to a Saudi pipeline across their territory.

Middle school assessment. Students will write a newspaper article reporting on the settlement of the Saudi–Yemeni border dispute by the Treaty of Jeddah, which was agreed to in June 2000. The article, written from either a Saudi or a Yemeni perspective, should describe the new boundary established by the treaty and analyze underlying physiographic and cultural considerations that influenced the location of that boundary. Students will also prepare a map to go with the article.

High school assessment. Students will write a newspaper article reporting on the settlement of the Saudi–Yemeni border dispute by the Treaty of Jeddah, which was agreed to in June 2000. The article, written from either a Saudi or a Yemeni perspective, should describe the new boundary established by the treaty and analyze underlying physiographic and cultural considerations that influenced the location of that boundary. The article should also include information about historical factors that contributed to this 65-year-old boundary conflict. A map should accompany the article.

Extending the lesson

Challenge students to try the following:

- Negotiate with each other in the introductory activity to arrive at a mutually satisfactory boundary for the two new classrooms.
- Use the Internet to identify other areas of the world where international boundaries are in dispute.
- Research the events of World War I on the Arabian Peninsula. Create an ArcMap map document illustrating these events.
- Use ArcMap to compare the countries of the Arabian Peninsula by mapping and analyzing relevant economic and demographic data.

See the "Resources by Module" section of this book's Web site—www.esri.com/ourworldgiseducation —for print, media, and Internet resources on the topics of Saudi Arabia, Yemen, and the Treaty of Jeddah.

Answer key

Step 2: Identify countries that border the Arabian Peninsula

Q1. What countries border the Arabian Peninsula to the north? Jordan, Iraq, Kuwait

Step 3: Investigate the physical characteristics of the Arabian Peninsula

Q2. Is any part of the Arabian Peninsula mountainous? Yes

Q3. If so, where are the mountains located? Along the west (Red Sea) coast and the northeast (Gulf of Oman) coast.

Q4. Are there any parts of the Arabian Peninsula that do not have any water at all? If so, where are these regions? Yes. The south-central part of the peninsula has no permanent bodies of water or streams.

Q5. Do you see any relationship between landforms and the availability of water? Mountains and areas of higher elevation have more surface water.

Q6. Describe the bodies of water in terms of their connectedness or disconnectedness. The permanent bodies of water look like a maze of rivers and lakes.

Q7. How many millimeters equal 10 inches? 254 mm

Q8. Based on the amounts of rainfall displayed on the map, do you think there is much farming on the Arabian Peninsula? Explain. No. Most of the Arabian Peninsula is so dry that agriculture wouldn't be possible without an alternative source of water such as a river. (Egypt, for example, is just as dry, but the Nile River provides water for agriculture.)

Q9. Approximately what percentage of the Arabian Peninsula is desert? 70–90%

Q10. What is the approximate range of temperatures across the Arabian Peninsula during September through November? 14 to 30°C, 57 to 86° F

Q11. Which season is the hottest? Summer (June–August)

Q12. What is the approximate range of temperatures across the Arabian Peninsula during the hottest season? 18 to 40°C, 66 to 104° F

Q13. What relationship do you see between the Arabian Peninsula's ecozones and its patterns of landforms, precipitation, and temperature? The limited zones of temperate grassland on the Arabian Peninsula are found in the mountains, where there is more precipitation and milder temperatures.

Q14. List three observations for each physical characteristic in the table.

Physical characteristics	Observations
Landforms and bodies of water	The Arabian Peninsula has a narrow coastal plain along the Red Sea that is separated by a low mountain range from the rest of the peninsula. The highest mountains in this range are at the southern point of the peninsula where the Red Sea meets the Gulf of Aden. A second, smaller range of mountains is found on the northeastern point of the peninsula along the Gulf of Oman. There are few permanent bodies of water and many intermittent streams.
Climate	Most of the Arabian Peninsula is a desert because it gets less than 10 inches of rainfall per year. Winter temperatures are mild, but summer temperatures are extremely high.
Ecozones	Most of the Arabian Peninsula has a desert ecosystem, with temperate grasslands being the primary exception. These grasslands can be found in the mountains and higher elevations of the western part of the peninsula and in the mountains of the northeast.

Q15. In your opinion, which of the region's physical characteristics would be considered "valuable" in a boundary decision? Explain. Grassland ecosystems, areas with greater than 500 mm (about 20 inches) of annual precipitation, and areas with access to permanent bodies of water or springs. Students may also mention mountains and passes important for defense or trade.

Step 4: Investigate the population characteristics of the Arabian Peninsula

Q16. What is the principal agricultural activity on the peninsula? Nomadic herding

Q17. Based on what you now know about the physical characteristics of the region, why do you think the agricultural activity is so limited? Most of the region does not have sufficient water for farming. Livestock can be herded from place to place depending on the seasonal availability of water and pastureland.

Q18. How does Yemen compare to the rest of the Arabian Peninsula in population density? Southwestern Yemen has the largest area of relatively high population density (more than 50 people per km^2).

Q19. Describe the overall population density of the Arabian Peninsula. Most of the Arabian Peninsula has fewer than 25 people per km^2, and at least half of that area has a density of less than 1 person per km^2.

Q20. Speculate about the ways water is most commonly used at these springs and water holes. Answers will vary. Because of the prevalence of nomadic herding, a logical conclusion is that most of the springs and water holes are used to water livestock.

123

Q21. Use your answers from Q16–Q20 and analysis of the maps to list two observations for each population characteristic in the table.

Population characteristics	Observations
Agricultural activities	The principal agricultural activity on the Arabian Peninsula is nomadic herding. Farming exists on a very limited basis in desert oases, on irrigated lands along the Red Sea, and in the high elevations of southern Yemen.
Population density and distribution	The Arabian Peninsula is very sparsely populated overall. The largest concentration of people is found around Sana in Yemen.

Q22. If an international boundary were to be drawn across some part of the Arabian Peninsula, how would these population characteristics influence the perception of certain regions as being more valuable than others? Nomadic herders would place a high value on having access to sources of water and grazing land. The population density of a region is a direct reflection of the ability of land to support life. Areas with very low population density would be least valuable, and those with high population density would be most valuable.

Step 5: Investigate the Empty Quarter

Q23. List two observations on the physical characteristics of the Empty Quarter and two observations on its population characteristics.
Physical characteristics: It's a desert region with virtually no permanent bodies of water and less than 3 inches of rainfall per year.
Population characteristics: The region has no agricultural activity and is virtually uninhabited: most of it has less than one person per square kilometer, and about one third has no people at all. The lack of roads indicates minimal human presence.

Q24. What difficulties would an area like this present if an international boundary must cross it? Insufficient residential population to draw from for boundary enforcement. The lack of permanent landmarks (due to shifting sand dunes) makes the line difficult to mark and see. Nomadic herders would want easy access throughout the border area.

Q25. Write the map document's new name and location.
Document Example: ABC_Region5.mxd
Location Example: C:\Student\ABC

Step 6: Explore Saudi Arabia's southern boundaries

Q26. Are the boundaries what you expected them to be? Answers will vary.

Q27. Which boundary remained unsettled? The border between Saudi Arabia and Yemen

Q28. What does the area between the green and purple lines represent? It is claimed by both Saudi Arabia and Yemen—it is the disputed territory between these two countries.

Q29. What is the principal economic activity of the regions in dispute? Nomadic herding

Q30. Describe the population distribution in the disputed territory. The disputed territory is mostly uninhabited, with a density of less than 1 person per km². The only area with a higher concentration of people is the western part of the territory with 1–25 people per km².

Step 7: Draw the Saudi–Yemeni boundary

Q31. Does the red line go through any cities or towns? If yes, approximately how many does the boundary pass through? Answers will vary. The boundary passes through fewer than 10 villages but within a mile or two (approximately 1.25 km) of many more.

Q32. How would you decide on which side of the town to put the boundary? Remember, this decision would determine whether the residents of that village would be citizens of Saudi Arabia or Yemen. Answers will vary. One way is to survey villagers to find out whether they feel a closer affiliation with Yemen or Saudi Arabia. Such affiliations are often based on long-standing tribal traditions.

Q33. Does the new line seem to favor Yemen or Saudi Arabia? Explain. The new border seems to favor Yemen. It gained control of all the disputed territory and gained even more territory beyond the previous boundary.

Step 8: Add the maritime portion of the boundary

Q34. What body of water does the maritime boundary traverse? The Red Sea

Q35. How does the actual boundary established by the Treaty of Jeddah compare with the boundary you drew earlier? Answers will vary. In most cases, students will find that the Treaty of Jeddah gave more land to Yemen than they predicted.

Q36. Write three observations about the boundary line created by the Treaty of Jeddah. Possible answers: The new boundary increased Yemen's territory. Most of Yemen's new territory is land used by nomadic herders and desert. The border settlement probably did not have a significant impact on Yemen's overall population as most of the new territory is uninhabited or very sparsely settled.

Step 9: Define the pastoral area

Q37. How many miles is 20 kilometers? 12.428 miles

Q38. In which portion of the Saudi–Yemeni border will the pastoral area be most significant? Explain. In the western portion, where nomadic herding is found. The remainder of the new boundary is north of the nomadic herding areas.

Q39. Why do you think the Treaty of Jeddah created a pastoral area? Answers will vary, but students should understand that the establishment of a pastoral area recognizes that nomadic herding is incompatible with fixed and finite boundaries. The pastoral area represents a compromise between the need to clearly define the boundary between Saudi Arabia and Yemen and the reality that the border area is populated by people whose nomadic traditions include territory on both sides of that boundary.

Assessment rubrics

Middle school

Standard	Exemplary	Mastery	Introductory	Does not meet requirements
1 The student knows and understands how to use maps to analyze spatial distributions and patterns	Creates a detailed GIS map that shows the new boundary line and relevant physical or cultural characteristics	Creates a map showing the new boundary line and most relevant physical or cultural characteristics	Creates a map showing the new boundary line but does not include any relevant physical or cultural characteristics	Creates a map of the Middle East but does not focus on the boundary issue between Saudi Arabia and Yemen
4 The student knows and understands how physical processes shape places and how different human groups alter the characteristics of places	Writes a detailed description of the physical and cultural characteristics that will be affected by the boundary change and explains possible ramifications	Writes a description of the physical and cultural characteristics that will be affected by the boundary change	Writes a description of physical or cultural characteristics that will be affected by the boundary change	Describes physical or cultural characteristics of the region but does not explain how these things are affected by the boundary change
13 The student knows and understands the multiple territorial divisions of the student's own world	Describes the implications of the new boundary for the people living in the affected areas; includes quotes or stories from individuals in the area (real or fictional)	Describes the implications of the new boundary for the people living in the affected areas	Describes characteristics of people living in the affected region but does not relate it specifically to the boundary change	Provides little description of the characteristics of the people living in the affected areas
18 The student knows and understands how various points of view on geographic context influence plans for change	Writes a coherent news article on the boundary issue; includes perspectives of both Saudi Arabia Yemen	Writes a coherent news article from the perspective of either country involved in the boundary issue	Writes a news article on the boundary issue but does not offer the perspective of either country on the boundary issue	Writes an essay on the boundary issue that does not offer a geographic perspective and is not in the form of a news article

This is a four-point rubric based on the National Standards for Geographic Education. The mastery level meets the target objective for grades 5–8.

High school

Standard	Exemplary	Mastery	Introductory	Does not meet requirements
1 The student knows and understands how to use geographic representations and tools to analyze and explain geographic problems	Creates a detailed GIS map showing the new boundary line and relevant physical and cultural characteristics	Creates a map showing the new boundary line and most relevant physical or cultural characteristics	Creates a map showing the new boundary line and relevant physical or cultural characteristics	Creates a map showing the new boundary line but does not include any relevant physical or cultural characteristics
4 The student knows and understands the changing human and physical characteristics of places	Writes a detailed description of the physical and cultural characteristics that will be affected by the boundary change and explains possible ramifications	Writes a description of the physical and cultural characteristics that will be affected by the boundary change	Writes a description of physical or cultural characteristics that will be affected by the boundary change	Describes physical or cultural characteristics of the region but does not explain how these things are affected by the boundary change
13 The student knows and understands why and how cooperation and conflict are involved in shaping the distribution of social, political, and economic spaces on Earth at different scales	Describes the implications of the new boundary for the people living in the affected areas in relationship to social issues, politics, and the economy; includes quotes or stories from individuals in the area (real or fictional)	Describes the implications of the new boundary for the people living in the affected areas in relationship to social issues, politics, and the economy	Describes the implications of the new boundary for the people living in the affected areas	Describes characteristics of people living in the affected region but does not relate it specifically to the boundary change
18 The student knows and understands contemporary issues in the context of spatial and environmental perspectives	Writes two coherent news articles on the boundary issue, one from the perspective of Saudi Arabia and one from the perspective of Yemen; includes historical factors that contributed to the conflict	Writes a coherent news article from the perspective of either country involved in the boundary issue; includes historical factors that contributed to the conflict	Writes a news article on the boundary issue but does not offer the perspective of either country on the boundary issue; includes one or two historical factors	Writes an essay on the boundary issue that does not offer a geographic perspective and is not in the form of a news article

This is a four-point rubric based on the National Standards for Geographic Education. The mastery level meets the target objective for grades 9–12.

Module 5: Lesson 2

127

MODULE 6

Economics

Lesson 1: The wealth of nations: A global perspective

Students will look at three modes of economic production—agriculture, industry, and services—as the initial criteria for determining a country's level of economic development. They will add layers of data representing additional economic indicators—energy use and gross domestic product (GDP) per capita—and draw their own conclusions on how economically developed certain countries are.

Lesson 2: Share and share alike: A regional investigation of North America

Students will explore trade between the three countries participating in the North American Free Trade Agreement (NAFTA): Canada, Mexico, and the United States. They will look at exports from each of the NAFTA countries for the past 16 years and use this information to identify trends and to assess NAFTA's effectiveness. They will create a layout containing a map and graphs that support their opinions.

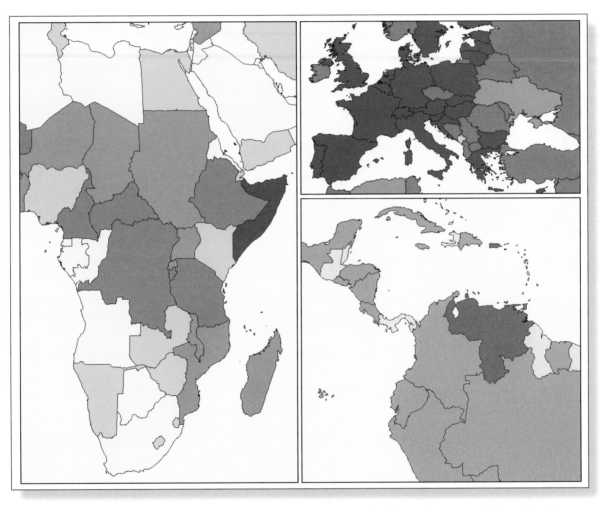

Analyzing world maps of economic output (GDP) points out relationships among the three main sectors: agriculture (green), services (red), and industry (blue).

The wealth of nations

A global perspective

Lesson overview

Students will look at three modes of economic production—agriculture, industry, and services—as the initial criteria for determining a country's level of economic development. They will add layers of data representing additional economic indicators—energy use and gross domestic product (GDP) per capita—and draw their own conclusions on how economically developed certain countries are.

Estimated time

Two 45-minute class periods

Materials

The student worksheet files can be found on the Data and Resources CD. Install the teacher resources folder on your computer to access them.

Location: OurWorld_teacher\Module6\Lesson1
- Student PDF: M6L1_student.pdf
- Student answer sheet: M6L1_student_answer_sheet.doc
- Student assessments: M6L1_assessment.pdf

Objectives

After completing this lesson, a student is able to do the following:

- Define the three sectors traditionally used to determine the level of economic development
- Evaluate production by sector as a suitable measurement of economic development
- Understand additional economic indicators of economic development: energy consumption and GDP per capita
- Define "developed" and "developing"

GIS tools and functions

▢ Display a map layout

▢ Zoom the layout to view the whole page

🔍 Zoom in on the map

▶ Activate a data frame in a layout

🔭 Find a specific feature on the map and identify it

🌐 Zoom to the full map extent

⬇ Add a layer to the map

- Add a new data frame to a layout
- Symbolize data using graduated colors
- Manually change labels on a legend
- Build a query to exclude specific attribute values from classification
- Display excluded values in a No Data class

National Geography Standards

Standard	Middle school	High school
1 How to use maps and other geographic representations, tools, and technologies to acquire, process, and report information from a spatial perspective	The student understands how to make and use maps, globes, graphs, charts, models, and databases to analyze spatial distributions and patterns	The student understands how to use geographic representations and tools to analyze, explain, and solve geographic problems
11 The patterns and networks of economic interdependence on Earth's surface	The student understands ways to classify economic activity	The student understands the classification, characteristics, and spatial distribution of economic systems
18 How to apply geography to interpret the present and plan for the future	The student understands how varying points of view about geographic context influence plans for change	The student understands how to use geographic knowledge, skills, and perspectives to analyze problems and make decisions

M ● ● ● ● ● ● ○
L ● ○

Teaching the lesson

Introducing the lesson

Begin with a discussion of the three economic production sectors: agriculture, industry, and services. Explain to students that economists generally rate a country as developing or developed by how much of its GDP is generated by agriculture, industry, and services. Generally, countries with a high percentage of GDP in agriculture (whether it be subsistence, commercial, or another type) are placed in the category of "developing."

Ask the following questions to elicit knowledge, beliefs, or ideas your students may already have about countries around the world:

* Name several countries that have a high percentage of GDP in agriculture
* Name a country that has a high percentage of GDP in services
* Which countries do we generally think of as highly industrialized (with high percentages of GDP in industry)?
* What other factors might be helpful in determining whether a country is developed or developing?

Student activity

We recommend that you complete the activity yourself before presenting the lesson in class. Doing so will allow you to modify the activity to accommodate the specific needs of your students. If they will not be working on individual computers, be sure to explain any necessary modifications.

In order for students to complete the assessment, they must save their project at the end of the activity. You might want to let them know in advance where to save it.

The following are things to look for while students are working on this activity:

* Are the students using a variety of tools to obtain the information they need?
* Are students experiencing any difficulty symbolizing the energy use and GDP per capita layers?

Concluding the lesson

After the students complete the lesson and the assessment, discuss their findings. If you have time, you can have each group share findings on an overhead projector and explain how they came to their conclusions. Students can also take turns presenting the thematic maps they have created, either in printed format or on a computer projection device from the front of the room. Conclude the lesson by asking the students to explain which factors they feel are most important in deciding if a country is developed or developing and to provide support for their choices. Do they feel this two-class system (developed and developing) is adequate? What alternative or additional classes can they suggest to describe a country's economic status?

Assessment. Students are asked to choose one country they believe is developed and another they believe is developing. Students will use the thematic mapping skills they have learned in this lesson to map additional indicators and draw conclusions about the two countries. They will write an essay on measuring economic development.

Extending the lesson

Challenge students to try the following:

- Look at jobs advertised in the local newspaper and classify them as agriculture, industry, or services. Which type(s) of employment predominates in your community?
- Research job openings in two counties and compare the counties in terms of percentages of jobs in agriculture, industry, and services.
- Compute GDP per capita for the top 10 trading partners of the United States and map it.
- Analyze standard-of-living and economic indicators for any country and present findings orally.

See the "Resources by Module section" of this book's Web site—www.esri.com/ourworldgiseducation— for print, media, and Internet resources on economic development indicators and related topics.

Answer key

Step 2: Examine the legends and patterns of the maps

Q1.   What do the darkest colors represent? Highest percentage of GDP in that sector

Q2.   What do the lightest colors represent? Lowest percentage of GDP in that sector

Q3.   What is the range of percentages for the darkest color of the agriculture layer in the table of contents? 60.1-100%

Q4.   Most countries with >40% of GDP in agriculture are located on which continent? Africa

Q5.   On which continents do all countries have ≤40% of GDP in agriculture? North America, South America, Europe, Australia

Q6.   On which continents do all countries have >40% of GDP in services? North America, South America, Australia

Q7.   Most countries with ≤40% of GDP in services are located on which continents? Africa and Asia

Q8.   What relationship, if any, do you see between the agriculture and services maps? They are generally opposite: where one is high, the other is low.

Q9.   Which continent has the most countries with ≤20% of GDP in industry? Africa

Q10.   According to the three economic sector maps and your answers in Q4–Q9, where are most of the developing countries located? Africa

Step 3: Analyze data for Ecuador

Q11.   What percentage of Ecuador's GDP is in agriculture? 6.3%

Q12.   What percentage of Ecuador's GDP is in industry? 33.5%

Q13.   What percentage of Ecuador's GDP is in services? 60.2%

Q14.   Would you classify Ecuador as a developed or developing country? Explain. Developed. Students should support their answer with the data.

Q15.   Place a check mark for Ecuador under Developed or Developing in the table. See the table below.

Country	GDP in the following sector			Economic status	
	Agriculture	Industry	Services	Developing	Developed
Ecuador	Low	Moderate	High		✓
Saudi Arabia	Low	High	Moderate		✓
Australia	Low	Moderate	High		✓
South Korea	Low	High	High		✓
Ukraine	Low	High	Moderate		✓
Congo DRC	High	Low	Moderate	✓	

Step 4: Analyze data for other countries

Q16. Record the category for Saudi Arabia's percentage of GDP in agriculture in the table in Q15. See the table on the previous page.

Q17. Record the category for Saudi Arabia's percentage of GDP in industry in the table in Q15. See the table on the previous page.

Q18. Record the category for Saudi Arabia's percentage of GDP in services in the table in Q15. See the table on previous page.

Q19. Place a check mark for Saudi Arabia under Developed or Developing in the table in Q15. See the table on previous page.

Q20. Fill in the information for Australia, South Korea, Ukraine, and People's Democratic Republic of Congo (Congo, DRC) in the table in Q15. See the table on previous page.

Step 7: Analyze GDP per capita and energy use data

Q21. What is Ecuador's GDP per capita? $600–6,400 (the lowest category)

Q22. What is Ecuador's annual amount of energy use? 0–2 quadrillion BTUs (the lowest category)

Q23. Based on this new information, should Ecuador be classified as a developing or developed country? Developing

Q24. Why does energy use increase when a country develops? Energy is used more when countries build infrastructure and establish manufacturing plants than when the primary mode of economic production is agriculture. Industry- and service-oriented production consumes more energy than agriculture.

Q25. Complete the following table. Categorize countries with <$16,000 as low GDP/capita and countries with <11 quadrillion BTUs as low on energy use.

Country	GDP per capita ($)	Energy use (quadrillion BTUs)	Developed or developing	Is this a change from your earlier classification?
Ecuador	Low	Low	Developing	Yes
Saudi Arabia	Low	Low	Developing	Yes
Australia	High	Low	Developed	No
Republic of Korea (South Korea)	Moderate	Low	Developed	No
Ukraine	Low	Low	Developing	Yes
Congo DRC	Low	Low	Developing	No

M ● ● ● ● ● ● ○
L ● ○

Q26. Name a country that you classified in Q15 as developed on the basis of economic sector data and in Q25 as developing on the basis of GDP per capita and energy use data. Possible answers: Ukraine, Saudi Arabia, Ecuador

Q27. Based on the data you collected on these six countries, do you feel that the economic sector criteria are good indicators of a country's economic status? Explain your answer. Answers will vary. Economic sector criteria appear to be good indicators of developing or developed status in some cases, but not all. Students should realize that many factors make up a country's economic status, and as different factors are included as criteria, a country's classification may change. Although Ukraine, Saudi Arabia, and Ecuador have high percentages of their GDPs in industry and services, they have low GDPs per capita and low energy use. Inclusion of these additional factors makes the three countries lean toward a "developing" classification.

Q28. Write the new name you gave the map document and where you saved it.
Document Example: ABC_Global6.mxd
Location Example: C:\Student\ABC

Assessment rubrics

Middle school

	Standard	Exemplary	Mastery	Introductory	Does not meet requirements
1	The student understands how to use maps, globes, graphs, charts, models, and databases to analyze spatial distributions and patterns	Uses GIS to analyze economic data by creating at least seven thematic maps that compare and contrast the different economic indicators for two countries; uses additional data from outside sources	Uses GIS to analyze economic data by creating seven thematic maps that compare and contrast the different economic indicators for two countries	Uses GIS to analyze economic data by creating five or six thematic maps that compare and contrast the economic indicators for two countries	Uses GIS to create four or fewer thematic maps based on economic data for one or two countries
11	The student understands ways to classify economic activity	Clearly describes the three economic sectors and provides an example of a country where one sector predominates; creates original, detailed, accurate definitions of "developed" and "developing"; provides ample evidence for the definition	Clearly describes the three economic sectors and creates accurate and original definitions of "developed" and "developing"	Describes the three economic sectors and attempts to create original definitions of "developed" and "developing"	Has difficulty describing the three economic sectors and does not attempt to create original definitions for "developed" and "developing"
18	The student understands how varying points of view about geographic context influence plans for change	Maps all listed indicators; compares own findings with the predefined economic statuses of the selected countries; identifies any inconsistencies in the findings and either accepts or rejects his/her hypothesis.	Maps eight indicators; compares own findings with the predefined economic statuses of the selected countries; identifies any inconsistencies in the findings and either accepts or rejects his/her hypothesis	Uses indicators from Global6 map document; compares own findings with the predefined economic statuses of the selected countries	Lists own ideas on the economic status of a selected country (or countries) but does not draw any comparisons with predefined economic status

This is a four-point rubric based on the National Standards for Geographic Education. The mastery level meets the target objective for grades 5–8.

High school

Standard	Exemplary	Mastery	Introductory	Does not meet requirements
1 The student understands how to use geographic representations and tools to analyze, explain, and solve geographic problems	Uses GIS to analyze economic data by creating at least seven thematic maps that compare and contrast the different economic indicators for two countries; uses additional data from outside sources	Uses GIS to analyze economic data by creating seven thematic maps that compare and contrast the different economic indicators for two countries	Uses GIS to analyze economic data by creating five or six thematic maps that compare and contrast the economic indicators for two countries	Uses GIS to create four or fewer thematic maps based on economic data for one or two countries
11 The student understands the classification, characteristics, and spatial distribution of economic systems	Clearly describes the three economic sectors and explains the relationship between them; creates original, detailed, accurate definitions of "developed" and "developing"; provides ample evidence and examples for each	Clearly describes the three economic sectors and explains the relationship between them; creates original, detailed definitions of "developed" and "developing"; provides some evidence and examples for each	Describes the three economic sectors and attempts to create original definitions of "developed" and "developing"	Has difficulty describing the three economic sectors and does not attempt to create original definitions for "developed" and "developing"
18 The student understands how to use geographic knowledge, skills, and perspectives to analyze problems and make decisions	Compares own findings with the predefined economic statuses of the selected countries; identifies any inconsistencies in the findings and either accepts or rejects his/her hypothesis; refines original hypothesis based on the findings; predicts future changes in the economic statuses of the countries, giving specific examples	Compares own findings with the predefined economic statuses of the selected countries; identifies any inconsistencies in the findings and either accepts or rejects his/her hypothesis; predicts future changes in the economic statuses of the countries	Compares own findings with the predefined economic statuses of the selected countries; attempts to predict future changes in economic status but does not provide enough evidence	Lists own ideas on the economic status of a selected country (or countries) but does not draw any comparisons with predefined economic status

This is a four-point rubric based on the National Standards for Geographic Education. The mastery level meets the target objective for grades 9–12.

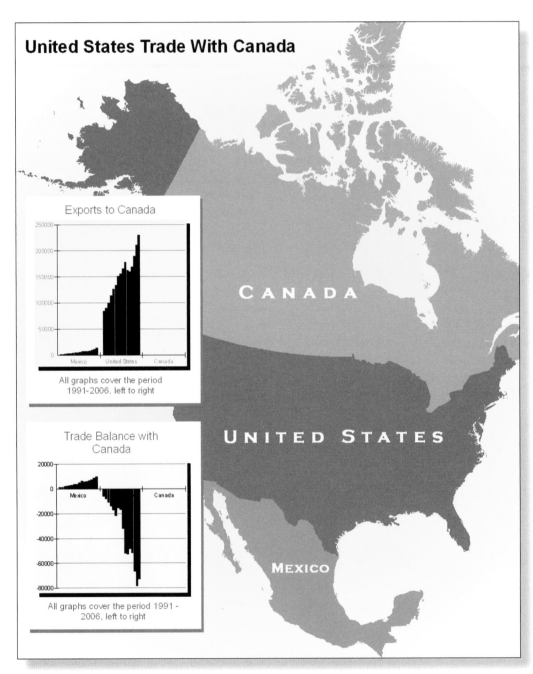

This layout explores the effects of the North American Free Trade Agreement (NAFTA).

Share and share alike

A regional investigation of North America

Lesson overview

Students will explore trade between the three countries participating in the North American Free Trade Agreement (NAFTA): Canada, Mexico, and the United States. They will look at exports from each of the NAFTA countries for the past 16 years and use this information to identify trends and to assess NAFTA's effectiveness. They will create a layout containing a map and graphs that support their opinions.

Estimated time

Three 45-minute class periods

Materials

The student worksheet files can be found on the Data and Resources CD. Install the teacher resources folder on your computer to access them.

Location: OurWorld_teacher\Module6\Lesson2
* Student PDF: M6L2_student.pdf
* Student answer sheet: M6L2_student_answer_sheet.doc
* Student assessments: M6L2_assessment.pdf

Objectives

After completing this lesson, a student is able to do the following:

* Explain the concept of trade balance
* Evaluate the effectiveness of NAFTA
* Create a layout displaying the results of the student's research

GIS tools and functions

- Display attributes for a specific feature
- Select a feature on the map
- Clear the selected features
- Switch from Data View to Layout View
- Zoom the layout to view the whole page
- Zoom out a fixed amount on the map
- Move elements in a layout
- Move the map to bring a different portion of it into view
- Add text to a layout

- Open the attribute table for a layer
- Relate a table to a layer attribute table
- Update a related table to reflect a change in selected features
- Open a graph and display data for a specific map feature
- Add graphs to a layout
- Print a layout

National Geography Standards

Standard	Middle school	High school
1 How to use maps and other geographic representations, tools, and technologies to acquire, process, and report information from a spatial perspective	The student understands how to make and use maps, globes, graphs, charts, models, and databases to analyze spatial distributions and patterns	The student understands how to use geographic representations and tools to analyze, explain, and solve geographic problems
11 The patterns and networks of economic interdependence on Earth's surface	The student understands the basis for global interdependence	The student understands the increasing economic interdependence of the world's countries
13 How the forces of cooperation and conflict among people influence the division and control of Earth's surface	The student understands how cooperation and conflict among people contribute to economic and social divisions of Earth's surface	The student understands why and how cooperation and conflict are involved in shaping the distribution of social, political, and economic spaces on Earth at different scales

Teaching the lesson

This lesson is intended to supplement a unit on NAFTA. For background, see the "Resources by Module" section of this book's Web site—www.esri.com/ourworldgiseducation. In addition to background information on NAFTA, students should have a working knowledge of basic economics. Make sure they are familiar with the terms trade, imports, exports, trade balance, and tariffs. Basic graph-reading skills will also be essential.

Introducing the lesson

Begin with a discussion of the NAFTA Objectives supplement. Ask students to explain the objectives of NAFTA in their own words. This exercise could be done as a class activity on the board or in groups with several sets of goals being developed. You may want to save the goals so students can refer to them when they evaluate NAFTA near the end of the lesson and in the assessment.

When you are satisfied that students have a basic understanding of NAFTA, discuss the flow of goods into and out of a country and the concept of trade balance. Introduce the following formula for trade balance:

Exports – Imports = Trade Balance
When exports exceed imports, you have a Trade Surplus.
When imports exceed exports, you have a Trade Deficit.

Ask students to predict trade balances for each of the NAFTA countries.

Student activity

We recommend that you complete the activity yourself before presenting the lesson in class. Doing so will allow you to modify the activity to accommodate the specific needs of your students. If they will not be working on individual computers, be sure to explain any necessary modifications.

Explain that in this activity students will create graphs and maps to explore and report on the patterns of trade over the past 16 years between the NAFTA countries. Note that the first three years of data (1991–1993) are prior to the inception of NAFTA. Students will use the graphs to identify whether NAFTA has achieved its goals as outlined in the objectives discussed in class.

The following are things to look for while the students are working on this activity:

* Are students using a variety of tools?
* Are they answering the questions?
* Do they need help with the lesson's vocabulary?

In Q33 students are asked to save their work according to the teacher's instructions. This is a good spot to stop the class for the day and to pick up the activity the next day. You might want to let students know in advance how to rename their project and where to save it.

Concluding the lesson

Have students present to the class their layouts and their initial predictions. Ask them to describe the trade balances and whether they think NAFTA has succeeded in achieving its goals. The students should also take a position on whether or not NAFTA should remain in place. Tally up the results and report to the class what their combined opinion suggests.

Middle school assessment. Students will create and present layouts that illustrate the changing trade balances between Mexico, Canada, and the United States, using maps they have made, graphs, text, and graphics. They will assess whether NAFTA has achieved its goals and take a position on whether it should remain in place, using information in their layouts to support their positions.

High school assessment. Students will create and present layouts that illustrate the changing trade balances between Mexico, Canada, and the United States, using maps they have made, graphs, text, and graphics. They will assess whether NAFTA has achieved its goals and take a position on whether it should remain in place, using information in their layouts to support their positions.

They will also write a paragraph describing how they would change NAFTA to improve or enhance trade for all three countries in the future.

Extending the lesson

Challenge students to try the following:

- Research the history of NAFTA.
- Acquire similar data for the European Union, Organization of Petroleum Exporting Countries (OPEC), and the Organization of American States, and compare and contrast the outcomes of NAFTA and the unification of Europe.
- Research the Central American Free Trade Agreement (CAFTA) and compare it with NAFTA.
- Map the movements of commodities from the NAFTA countries in which they were produced to countries that bought them.
- Go to the Statistics Canada, U.S. Census Bureau, and Instituto Nacional de Estadística Geografía e Informática (INEGI) Web sites to obtain information on total international exports and map it.
- Research employment trends since the advent of NAFTA. Have the NAFTA countries experienced increased employment or unemployment since 1994?

See the "Resources by Module" section of this book's Web site—www.esri.com/ourworldgiseducation—for print, media, and Internet resources on the topics of economics, trade, and NAFTA.

Answer key

Step 2: Examine the map and attribute table

Q1.   For which years does the layer contain data? 1991 to 2006

Q2.   How many attributes are there for each year? Six

Q3.   What was the value of goods and services exported from Canada to the United States in 1991? $91,064,000,000

Q4.   What is the name of the table? Attributes of NAFTA Countries

Q5.   How many rows are there for each country on the map? One

Step 3: Relate another table to the layer table

Q6.   How many rows are there for each country? Four

Q7.   What do the rows for Canada say under Item? Total Exports to United States, Trade Balance with United States, Total Exports to Mexico, Trade Balance with Mexico

Q8.   What are the general types (or items) of information listed in the table? Total exports and trade balances for each of the NAFTA countries

Q9.   How many years of data are represented in the table? 16

Q10. What happens in the Attributes of NAFTA_Trading_Statistics table? The rows for the selected country (United States) become highlighted.

Q11. What happens in the two tables and the map? Canada is highlighted (blue) on the map and in the NAFTA Countries table. Nothing changes in the NAFTA_Trading_Statistics table; the rows for the United States remain highlighted.

Q12. What happens in the two tables and the map? The Mexico row that was clicked in the NAFTA_Trading_Statistics table becomes highlighted, but Canada remains selected in the NAFTA Countries table and on the map.

Q13. What have you observed about the way the NAFTA_Trading_Statistics table is tied to the NAFTA Countries table and map layer? Changes made in one table are not automatically reflected in the other table or on the map. The Related Tables function can be used to make the same changes in the second table.

Step 4: Examine export graphs

Q14. Which country exported more goods and services to Canada: Mexico or the United States? United States

Q15. Why is the graph empty in the space for Canada? A country doesn't export goods to itself.

Q16. What happened to the graph? Only the data for Mexico (the selected country) is displayed.

Q17. How many years of data are represented on the graph? 16

Q18. What year does the first bar on the left represent? 1991

Q19. Compare the numbers on the y-axis with those in the two tables. Are the numbers on the graph in thousands, millions, or billions of dollars? Millions

Q20. Looking at the graph, how would you describe the trend of Mexican exports to Canada over the 16-year period? The exports increased significantly.

Q21. What was the approximate value of Mexican exports to Canada in 1991? $2,300,000,000 ($2.3 billion). In 2006? $14,000,000,000 ($14 billion)

Q22. Approximately how many times greater is the 2006 amount of Mexican exports to Canada than the 1991 amount? 6

Q23. How would you describe the trend of Mexican exports to the United States over the 16-year period? They increased significantly.

Q24. Approximately how many times greater is the 2006 amount of Mexican exports to the United States than the 1991 amount? 6 or 7

Step 5: Examine a trade balance graph

Q25. Did Mexico have a trade surplus or deficit with the United States for 1992? Deficit

Q26. What was the approximate value of the trade balance for 1992? $–5,000,000,000 ($–5 billion)

Q27. What was the first year that Mexico exported more to the United States than it imported from the United States? 1995

Q28. Describe the trend of Mexico's trade balance with the United States over the 16-year period. Small deficits in 1991–1994; significantly increasing surpluses in 1995–2006.

Q29. Did Canada have a deficit trade balance with the United States anytime during the 16-year period? No

Q30. In 1998, was Canada's trade balance with the United States greater, smaller, or about the same as Mexico's? About the same

Q31. In 2006, was Canada's trade balance with the United States greater, smaller, or about the same as Mexico's? Greater

Q32. What was the exact value of Canada's trade balance with the United States in 2006? $72,836,000,000 surplus

Q33. Write the map document's new name and location.
Document Example: ABC_Region6.mxd
Location Example: C:\Student\ABC

Step 6: Evaluate the effectiveness of NAFTA

Q34. Determine from the graphs the approximate value of exports in 2006 for each pair of countries in the table on the following page. Record the values in the middle column. Approximate answers are listed in the table.

Direction of export flow	Value of exports (million $)	Total volume between partners (million $)
United States to Mexico	135,000	335,000
Mexico to United States	200,000	
United States to Canada	230,000	530,000
Canada to United States	300,000	
Canada to Mexico	5,000	20,000
Mexico to Canada	15,000	

Q35. Add the export values together for each pair of countries (for example, exports from the United States to Mexico plus exports from Mexico to the United States). Record the totals in the last column in the table in Q34. See the table above.

Q36. Rank the trading partners by total volume of trade. Use 1 for the partners trading the most and 3 for the partners trading the least. United States–Mexico: 2 United States–Canada: 1 Canada–Mexico: 3

Q37. Did NAFTA have a positive (+), negative (–), or neutral (N) effect on trade volume between partner countries? United States–Mexico: + United States–Canada: + Canada–Mexico: +

Q38. Do you think that any one of these three countries benefited more than the other two from NAFTA? If so, which country? Explain your answer. Answers will vary because students are asked to speculate.

Q39. What country has a healthier trade balance with Canada: Mexico or the United States? Mexico

Q40. On what graph do you find a set of bars that looks like a mirror image of those for the U.S. trade balance with Canada? Trade Balance with U.S.

Q41. What country had the most dramatic change for the better in trade after NAFTA came into being in 1994? Mexico

Q42. Estimate U.S. trade deficits with Canada and Mexico for 2004 and 2006 and record them (in billion dollars) in the table below. Approximate answers are listed in the table.

Trading partner	U.S. trade deficit ($)	
	2004	2006
Mexico	–45 billion	–64 billion
Canada	–66 billion	–73 billion
Combined	–111 billion	–137 billion

Q43. Did the U.S. combined trade balance get better or worse between 2004 and 2006? Worse By how much? $26 billion

Assessment rubrics

Middle school

Standard	Exemplary	Mastery	Introductory	Does not meet requirements
1 The student understands how to make and use maps, globes, graphs, charts, models, and databases to analyze spatial distributions and patterns	Creates a GIS layout with eight or more components, including several maps and graphs that illustrate the effectiveness of NAFTA	Creates a GIS layout with eight components, including a map and two or more graphs	Creates a GIS layout with most of the eight specified components, including a map and at least one graph	Creates a GIS layout with some of the eight specified components, including a map
11 The student understands the basis for global interdependence	Layout shows a clear understanding of the concepts of balance of trade for NAFTA trading partners through multiple graphs and maps	Layout includes graphs and maps showing an understanding of the concept of balance of trade for NAFTA trading partners	Layout includes graphs and maps showing some understanding of the concept of balance of trade for NAFTA trading partners	Layout includes a graph and/or a map showing trade partners but does not illustrate an understanding of balance of trade
13 The student understands how cooperation and conflict among people contribute to economic and social divisions of Earth's surface	Presents a logical argument for his/her conclusions on the effectiveness of NAFTA, using a variety of resources to support his/her findings, including, but not limited to, data and maps	Presents a logical argument for his/her conclusions on the effectiveness of NAFTA using data and maps to support his/her ideas	Takes a position on the effectiveness of NAFTA and provides some data and maps as evidence	Identifies NAFTA participants but does not assess NAFTA's effectiveness

This is a four-point rubric based on the National Standards for Geographic Education. The mastery level meets the target objective for grades 5–8.

High school

Standard	Exemplary	Mastery	Introductory	Does not meet requirements
1 The student understands how to use geographic representations and tools to analyze, explain, and solve geographic problems	Creates a GIS layout with eight or more components, including several maps and graphs that illustrate the effectiveness of NAFTA	Creates a GIS layout with eight components, including a map and two or more graphs that illustrate the effectiveness of NAFTA	Creates a GIS layout with most of the eight specified components, including a map and at least one graph that illustrate the effectiveness of NAFTA	Creates a GIS layout with some of the eight specified components, including a map that illustrates the effectiveness of NAFTA
11 The student understands the increasing economic interdependence of the world's countries	Composes a brief essay proposing specific improvements to NAFTA that will benefit all three countries; provides examples	Composes a paragraph proposing specific improvements to NAFTA that will benefit all three countries	Composes a paragraph proposing improvements to NAFTA which may not benefit all parties	Lists proposed improvements to NAFTA but does not show how these improvements will enhance trade
13 The student understands why and how cooperation and conflict are involved in shaping the distribution of social, political, and economic spaces on Earth at different scales	Presents a logical argument for his/her conclusions on the effectiveness of NAFTA at the local, regional, and global levels using a variety of resources to support his/her findings, including, but not limited to, data and maps	Presents a logical argument for his/her conclusions on the effectiveness of NAFTA at the local, regional, and global levels using data and maps to support his/her ideas	Takes a position on the effectiveness of NAFTA at the local, regional, and global levels but does not provide a strong argument with ample evidence	Identifies NAFTA participants but does not assess NAFTA's effectiveness

This is a four-point rubric based on the National Standards for Geographic Education. The mastery level meets the target objective for grades 9–12.

Module 6: Lesson 2

MODULE 7

Forces of nature

Lesson 1: Water world: A global perspective

Students will investigate changes that might occur to the surface of the earth if the major ice sheets of Antarctica melted. They will begin their exploration at the South Pole by studying the physical geography of Antarctica. They will consider the consequences of projected changes on human structures, both physical and political. The assessment asks students to create an action plan for a major city that would be flooded in the event of a catastrophic polar meltdown.

Lesson 2: In the eye of the storm: A regional investigation of Central America

Students will study Hurricane Mitch, the deadliest storm of the twentieth century, and the havoc it wreaked on several Central American countries. They will analyze information about the storm itself and about the region before the storm, and they will consider the storm's consequences.

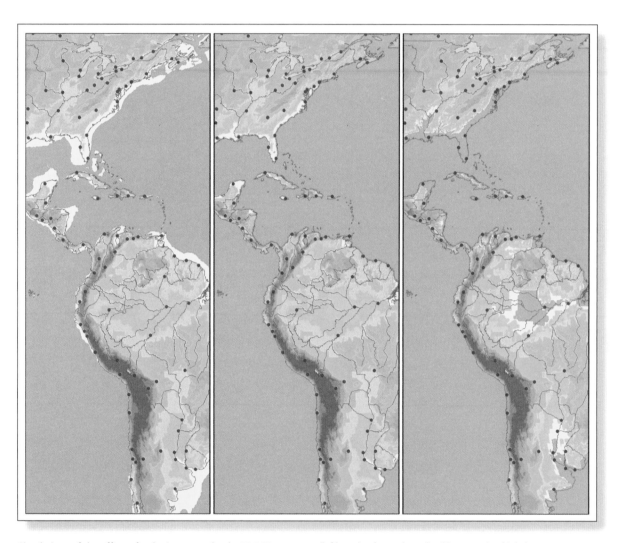

Simulations of the effect of polar ice on sea levels: 20,000 years ago (left), today (center), and a 50-meter rise (right).

Module 7: Lesson 1

Water world

A global perspective

Lesson overview

Students will investigate changes that might occur to the surface of the Earth if the major ice sheets of Antarctica melted. They will begin their exploration at the South Pole by studying the physical geography of Antarctica. They will consider the consequences of projected changes on human structures, both physical and political. The assessment asks students to create an action plan for a major city that would be flooded in the event of a catastrophic polar meltdown.

Estimated time

Two 45-minute class periods

Materials

The student worksheet files can be found on the Data and Resources CD. Install the teacher resources folder on your computer to access them.

Location: OurWorld_teacher\Module7\Lesson1
- Student PDF: M7L1_student.pdf
- Student answer sheet: M7L1_student_answer_sheet.doc
- Student assessments: M7L1_assessment.pdf

Objectives

After completing this lesson, a student is able to do the following:

- Manipulate map projections using GIS technology
- Compare locations on a map to photos and satellite imagery using GIS
- Analyze the impact on major human systems (such as transportation networks) that would be caused by the melting of various parts of the Antarctic ice sheet, causing a significant rise in sea level
- Predict how such a catastrophe might change the nature of cities and societies around the world, and propose ways to minimize danger and hardship

GIS tools and functions

  Zoom in and out of the map

 View a hyperlink to an image

 Add layers to the map

- Change the map projection for a data frame
- Activate a second data frame

National Geography Standards

Standard	Middle school	High school
1 How to use maps and other geographic representations, tools, and technologies to acquire, process, and report information from a spatial perspective	The student understands the relative advantages and disadvantages of using maps, globes, aerial and other photographs, satellite-produced images, and models to solve geographic problems	The student understands how to use geographic representations and tools to analyze, explain, and solve geographic problems
4 The physical and human characteristics of places	The student understands how different physical processes shape places	The student understands the changing physical and human characteristics of places
11 The patterns and networks of economic interdependence on Earth's surface	The student understands the basis for global interdependence and the reasons for the spatial patterns of economic activities	The student understands the increasing economic interdependence of the world's countries
18 How to apply geography to interpret the present and plan for the future	The student understands how the interaction of physical and human systems may shape present and future conditions on Earth	The student understands how to use geographic knowledge, skills, and perspectives to analyze problems and make decisions

Teaching the lesson

Introducing the lesson

Begin the lesson with a discussion of Antarctica. Use these questions as a guide:

- What is the climate of Antarctica like?
- What does the place look like?
- Are there any human settlements there?

After a brief discussion introducing the subject, share with your students some of the work that scientists have been doing in the region. You may want to have students explore some of the Internet sites associated with this lesson on the book's Web site. These resources provide information on the latest research into snow and ice melt in Antarctica and its impact on mean sea level.

M ● ● ● ● ● ● ●
L ● ○

In the activity, students will look at visual representations of the rising sea level and analyze current data about cities and important human structures. They will be challenged in the closing assessment to save a major city from rising floodwaters by using the data they gather in the course of their investigation.

Student activity

We recommend that you complete the activity yourself before presenting the lesson in class. Doing so will allow you to modify the activity to accommodate the specific needs of your students. If they will not be working on individual computers, be sure to explain any necessary modifications.

After investigating Antarctica, students will explore how changes in sea level could affect important human structures. If you have only one class period at computers, you can focus on map projections of Antarctica and stop before step 5 or you can begin at that point if you want to focus on the rise in sea level.

The following are things to look for while the students are working on this activity:

- Are the students using a variety of tools?
- Are the students answering the questions?
- Are the students beginning to ask their own questions?

If your students are having trouble getting the Hyperlink tool to work, try expanding the area you are allowed to click:

1. Click the Selections menu and click Options.
2. Type a bigger number in the Selection tolerance box. For example, change 3 to 6.
3. Click OK. Now the tool should be easier to use.

Concluding the lesson

Briefly discuss the observations students have made on the repercussions of rising sea levels. How similar or different are these observations?

Middle school assessment. Divide students into teams of three. Each team will select (or you can assign to them) a city from a list of those likely to be greatly affected by a 50-meter rise in sea level and create an action plan for relocating the city and its resources. They will focus on basic modes of transportation such as major roads and railways and may also address effects on utilities. Each team should have a leader, a cartographer, and a data expert with the following responsibilities:

- Leader—organizes the group and coordinates creation of the final product.
- Cartographer—primarily responsible for manipulating the GIS and printing any maps for the final product.
- Data expert—focuses on research and determines which data is best to use.

Results can be presented orally, in writing, or in a science-fair-type poster.

High school assessment. Divide students into teams of three. Each team will select (or you can assign to them) a city from a list of those likely to be greatly affected by a 50-meter rise in sea level. They will then create an action plan to relocate the city, shift the city's national and international roles to another city, adapt the city to its new environment, or develop another strategy. They must take into account transportation, utilities, economics, and trade relations in their solutions. Each team should have a leader, a cartographer, and a data expert. Results can be presented orally, in writing, or in a science-fair-type poster.

155

Extending the lesson

Challenge students to try the following:

- Research the affected cities and countries by obtaining additional data from outside sources such as the Internet.
- Create action plans for an entire country. One group could focus on political boundaries, another on transportation, another on export and trade. After each group presents its findings, the class can explore how the different plans would work together.
- Identify a city along the Mississippi River that frequently floods. Research and analyze its flood disaster plans. How could those plans be improved?

See the "Resources by Module" section of this book's Web site—www.esri.com/ourworldgiseducation— for print, media, and Internet resources on the topic of Antarctica and global warming.

Answer key

Step 2: Look at Antarctica

Q1. Do you think this map gives you a realistic representation of Antarctica? Explain. Answers will vary but should mention that the map of Antarctica is very skewed in size, shape, and distance.

Q2. Does this projection give you a better view of the South Pole region? Why or why not? No. The South Pole region is distorted in size.

Q3. Do any of these projections work well for viewing Antarctica? No. None of the projections represents Antarctica in a realistic way.

Step 5: Activate the Water World data frame

Q4. What significant differences do you see between current landmass outlines and those of 20,000 years ago? List at least three. Possible answers: Alaska was connected to Russia, Florida was much larger, Australia was connected to the islands of Indonesia.

Step 6: Analyze global sea levels that would result if Antarctic ice sheets melted

Q5. Record your general observations for each layer in the table below. Possible answers are listed in the table.

Sea level	Observations
Today	Country outlines match up perfectly with the shorelines.
Plus 5 meters	The change is not dramatic, but some coastal cities in the southern United States (e.g., Miami) will be under water.
Plus 50 meters	The change is dramatic. Most of Florida is under water, there is a large gap in the Amazon Basin in South America, and The Netherlands is almost entirely gone.
Total thaw (plus 73 meters)	Large portions of Australia, South America, and southeast United States are gone. Africa is the least affected.

Step 7: View changes in water levels

Q6. What kinds of changes do you see in the rivers and lakes? Provide a specific example. Answers will vary. A large lake in the north-central area is the result of the increased water level of the Amazon River. The Parana River in Argentina and the Amazon River are significantly shorter.

Q7. With a sea level increase of 50 meters, what kinds of consequences do you foresee for the major river ecosystems of South America? Provide a specific example. Answers will vary. The Amazon River may flood the rain forests of central South America.

Q8. Some inland areas around the globe are below sea level. One of them is in South America. Hypothesize how these low-lying areas were formed. Answers will vary. One explanation is that plate boundaries are drifting apart at that location. Another is that the land surrounding the Amazon Basin could have risen up due to tectonic activity.

Step 8: View changes in political boundaries

Q9. Predict possible consequences of the 50-meter rise in sea level to the populations living in the Southwest Asia (political disputes, trade and economic issues, transportation problems, etc.). Record those consequences in the first row of the table below. A possible answer is listed in the table below.

Region	Countries/areas affected	Possible consequences
Southwest Asia	Iraq	Boundary disputes over lost land.
Asia	Cambodia	Much of Cambodia is under water. Many people would migrate to neighboring countries.
Europe	Netherlands	Netherlands would be almost entirely submerged. Many people would migrate to neighboring countries.
Africa	Senegal, Guinea-Bissau	Large portions of Senegal and Guinea-Bissau would be submerged. Many people would migrate to neighboring countries.
Oceania	Australia	Major cities and economic areas on the coast would be lost. People would move inland.
North America	United States	Most major southern ports—Houston, New Orleans, Miami (along with most of Florida)—would be lost. People would move inland.
South America	Brazil	People living in coastal cities would need to migrate inland. The Brazilian economy could suffer from the loss of natural resources flooded by the invasion of the sea into the Amazon basin.

Q10. Record your predictions in the table above. Possible answers are listed in the table above.

Q11. List other possible layers of data you might want to analyze to study the impact of rising sea levels. Answers will vary. See the Assessment table "Data Sources" for a list of available data.

Assessment rubrics

Middle school

Standard	Exemplary	Mastery	Introductory	Does not meet requirements
1 The student understands the relative advantages and disadvantages of using maps, globes, aerial and other photographs, satellite-produced images, and models to solve geographic problems	Creates a series of GIS maps illustrating the effects of a 50-meter rise in sea level on the selected city and the proposed phases of the action plan	Creates two GIS or paper maps illustrating the effects of a 50-meter rise in sea level for the selected city	Creates one map of the selected city that does not adequately illustrate the effects of the 50-meter rise in sea level	Creates a map that does not focus on the selected city
4 The student understands how different physical processes shape places	Creates a clear and concise hypothesis on the impact of a 50-meter rise in sea level on a city; provides data and maps that support ideas	Creates a clear and concise hypothesis on the impact of a 50-meter rise in sea level on a city; provides a map that supports ideas but does not highlight supporting data	Identifies some effects of a significant rise in sea level on a city but does not create a formal hypothesis; provides some data or a basic map	Identifies some places that will be affected by a rise in sea level but does not identify changes that could occur
11 The student understands the basis for global interdependence and the reasons for the spatial patterns of economic activities	Analyzes the effects on trade and transportation routes for a flooded city; develops an action plan that relocates the city and provides details on plan implementation; the plan uses data and maps to support ideas and includes a time line	Analyzes the effects on trade and transportation routes for a flooded city; develops an action plan that includes a time line	Attempts to analyze appropriate data and creates an outline for a plan but does not provide details on implementation; time line is incomplete	Does not select appropriate datasets (such as transportation) and therefore has difficulty creating a plan or a time line
18 The student understands how the interaction of physical and human systems may shape present and future conditions on Earth	Develops a coherent argument that supports the relocation plan; uses a variety of data to support the findings and creates additional data and maps that illustrate changes to the human infrastructure	Develops a coherent argument that supports the relocation plan; uses a variety of data to support the findings	Creates a relocation plan for the selected city but does not offer a variety of data to support ideas	Relocates a selected city but does not formalize a plan on how the change will take place; uses little or no data to support ideas

This is a four-point rubric based on the National Standards for Geographic Education. The mastery level meets the target objective for grades 5–8.

High school

Standard	Exemplary	Mastery	Introductory	Does not meet requirements
1 The student understands how to use geographic representations and tools to analyze, explain, and solve geographic problems	Creates a series of GIS maps illustrating the effects of a 50-meter rise in sea level on the selected city and the proposed phases of the action plan	Creates two GIS or paper maps illustrating the effects of a 50-meter rise in sea level for the selected city	Creates one map of the selected city that does not adequately illustrate the effects of the 50-meter rise in sea level	Creates a map that does not focus on the selected city
4 The student understands the changing physical and human characteristics of places	Clearly describes how the characteristics of a city could change, should a 50-meter rise in sea level occur; includes data from the Assessment table "Data Source" to create an original map to illustrate the changes	Clearly describes how the characteristics of a city could change, should a 50-meter rise in sea level occur; includes data from the Assessment table "Data Source"	Describes how some characteristics of a city could change, should a 50-meter rise in sea level occur; provides little evidence from the data	Attempts to describe how some characteristics of a city could change with a 50-meter rise in sea level; does not provide any evidence from the data
11 The student understands the increasing economic interdependence of the world's countries	Provides specific examples of how the loss of the selected city would affect the global marketplace; provides an economic action plan to prevent this loss from occurring	Identifies how the global economic infrastructure will be altered with a 50-meter rise in sea level, using the perspective of the selected city	Identifies how the selected city will be affected economically but does not make a connection to the global marketplace	Attempts to identify how the selected city will be affected economically but does not make a connection to the global marketplace
18 The student understands how to use geographic knowledge, skills, and perspectives to analyze problems and make decisions	Develops a detailed action plan to sustain the selected city in the event of a 50-meter rise in sea level; creates original GIS maps with a variety of data sources to support the ideas in the plan	Develops a detailed action plan to sustain the selected city in the event of a 50-meter rise in sea level; uses a variety of data sources to support the ideas in the plan	Develops a plan to sustain the selected city in the event of a 50-meter rise in sea level; provides little or no data to support ideas	Creates an outline of a plan to sustain the selected city in the event of a 50-meter rise in sea level; provides little or no data to support ideas

This is a four-point rubric based on the National Standards for Geographic Education. The mastery level meets the target objective for grades 9–12.

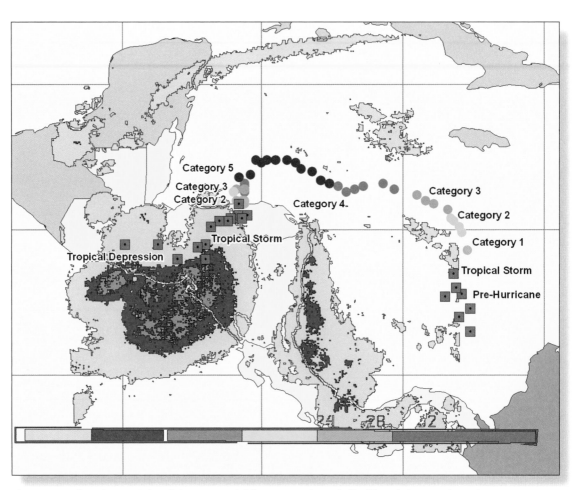

The progress of Hurricane Mitch and the storm's rainfall pattern at the end of its life cycle in Central America.

Module 7: Lesson 2

In the eye of the storm
A regional investigation of Central America

Lesson overview
Students will study Hurricane Mitch, the deadliest storm of the twentieth century, and the havoc it wreaked on several Central American countries. They will analyze information about the storm itself and about the region before the storm, and they will consider the storm's consequences.

Estimated time
Three 45-minute class periods

Materials
The student worksheet files can be found on the Data and Resources CD. Install the teacher resources folder on your computer to access them.

Location: OurWorld_teacher\Module7\Lesson2
- Student PDF: M7L2_student.pdf
- Student answer sheet: M7L2_student_answer_sheet.doc
- Student assessments: M7L2_assessment.pdf

Objectives
After completing this lesson, a student is able to do the following:

- Analyze regional features vulnerable to natural disasters
- Follow the development and impact of a major hurricane
- Integrate satellite imagery of a hurricane with a plot of the hurricane's path
- Develop a disaster relief plan

GIS tools and functions

- Identify the attributes of a feature
- Add layer files and images to a map
- Zoom in on a map
- Move the map to bring a different portion of it into view
- Measure distance on a map

- Zoom to the geographic extent of a layer
- Label features using overlapping labels
- Turn labels on and off
- Change a layer's symbol
- Activate and expand a second data frame
- Open an attribute table
- Sort a field in a table in ascending order
- Show only selected records in a table
- Show all records in a table
- Rearrange layers in the table of contents
- Hide or show a layer's legend in the table of contents

National Geography Standards

Standard	Middle school	High school
1 How to use maps and other geographic representations, tools, and technologies to acquire, process, and report information from a spatial perspective	The student understands the characteristics, functions, and applications of maps, globes, aerial and other photographs, satellite-produced images, and models to solve geographic problems	The student understands how to use technologies to represent and interpret Earth's physical and human systems
7 The physical processes that shape the patterns of Earth's surface	The student understands how to predict the consequences of physical processes on Earth's surface	The student understands the spatial variation in the consequences of physical processes across Earth's surface
15 How physical systems affect human systems	The student understands how natural hazards affect human activities	The student understands strategies of response to constraints and stresses placed on human systems by the physical environment, and how humans perceive and react to natural hazards
18 How to apply geography to interpret the present and plan for the future	The student understands how to apply the geographic point of view to solve social and environmental problems by making geographically informed decisions	The student understands how to use geographic knowledge, skills, and perspectives to analyze problems and make decisions

Teaching the lesson

Introducing the lesson

Begin the lesson by discussing the weather hazards of your own region. Use the following questions as a guide:

- What weather hazards are specific to our hometown or region?
- When do they typically occur—year-round or in a particular season?
- What are some characteristics of these phenomena?
- How do you prepare for one? You may want to review school procedures or have the students share plans they have from home.

After the discussion, tell your class that in this activity they will be studying the impact of Hurricane Mitch on a large area of Central America. They will explore characteristics of the storm and how it affected the region. In the final assessment, they will be part of a special team developing an action plan for dealing with the devastation the storm caused.

Student activity

We recommend that you complete the activity yourself before presenting the lesson in class. Doing so will allow you to modify the activity to accommodate the specific needs of your students. If they will not be working on individual computers, be sure to explain any necessary modifications.

The lesson provides a stopping point at Q8, which asks students to save the map document under a new name. You might want to tell students in advance how to rename the document and where to save it.

Because the image files that will be added to the map document are large, you may not want students to have individual copies of the module 7 folder. You can have students access the data in the central class folder instead.

Students will need to refer to their saved map documents to complete the assessment.

The following are things to look for while the students are working on this activity:

- Are the students using a variety of tools?
- Are the students answering the questions?
- Are the students beginning to ask their own questions?

Concluding the lesson

Have students share their findings, either in small groups or as a class. They should have a basic understanding of the region and the effect of the storm in terms of rainfall amounts, wind speeds, and so forth. Ask your students which parts of the region suffered the most damage and why. Encourage students to do their own research while completing the assessment. The U.S. Geological Survey has a large amount of data at http://mitchnts1.cr.usgs.gov on the impact of Hurricane Mitch and disaster recovery efforts in the affected countries. Allow class time for each team to meet and plan how they will complete the assessment.

Assessment. Students are asked to predict damage from Hurricane Mitch and to develop a disaster relief plan. Assign students to teams of three or four. Each team should have a leader, a cartographer, and a data expert. Teams of four can also have a multimedia specialist responsible for creating a presentation of the team's findings. Assign each team to deal with either flood hazards or volcano and landslide hazards in a particular country. Students can present their results orally or in writing.

Extending the lesson

Challenge students to do the following:

- Create a disaster relief plan for your city using a local natural hazard as the potential threat. Research past disasters in your city or region to find out how they affected the local area.
- Watch excerpts from the films *The Perfect Storm* or *Twister* and discuss how these fictional storms relate to real ones.
- Do a book report on *Isaac's Storm*, by Erik Larson, and compare the Galveston storm of 1900 to Hurricane Mitch.

See the "Resources by Module" section of this book's Web site—www.esri.com/ourworldgiseducation —for print, media, and Internet resources on Hurricane Mitch, Central America, and tropical storms.

Answer key

Step 2: Identify the capital cities of Central America

Q1. Record the capitals of Central American countries in the table below. Answers are listed in the table.

Central America Prior to Hurricane Mitch

Country	Populated places		Transportation				Precipitation	Agricultural use
	Capital	Distribution	Roads	Railroads	Airports			
Belize	Belmopan	Throughout the country	Sparse network	None	One near the coast	Primarily 1,401–2,800 mm	Primarily forest, with some irrigated land and little cropland
Guatemala	Guatemala	Throughout the country, but more concentrated near the Pacific Coast	Well developed network, especially in the south	Primarily in the south	Several distributed throughout the country	Full range, from <1,000 to >4,000 mm	About 1/2 forest mixed with irrigated and nonirrigated land, cropland, and forested wetlands along the south coast
Honduras	Tegucigalpa	Concentrated in the west	Well developed in the west	Along the northern coast	Along the northern coast and in the south near the capital; two off shore	From <1,000 mm in the center to >2,800 mm on the Caribbean Coast	1/4 forest, some cropland, some grazing land, 1/2 nonirrigated land
El Salvador	San Salvador	Throughout the country, but more concentrated around the capital	Well developed	Well developed	Near the capital and in the east	Primarily 1,401–2,800 mm	Primarily cropland with some grazing, forested wetlands along the coast, little forest
Nicaragua	Managua	Concentrated near the Pacific Coast and around the capital	Well developed in most of the country; less developed near the Caribbean Coast	Sparse; near capital	On the east and west coast near the capital; none in the center of the country; two off shore (actually on Columbian territory)	From <1,000 mm in the northwest to >4,000 mm along the Caribbean Coast	1/3 forest, some cropland and nonirrigated land, grazing land, some forested wetland along the east coast
Costa Rica	San Jose	Concentrated near the Pacific Coast and around the capital	Well developed along the Pacific Coast	Through center of the country and capital	Many airports; located near the coasts and the capital	Mostly 2,801–5,600 mm along the coastlines, with the exception of <1,400 mm in the northwest	Half cropland and grazing land, and 1/4 forested land
Panama	Panama	Concentrated in the west near the Pacific Coast; mostly unpopulated in the east	Well developed along the Pacific Coast west of the capital	Sparse	Clustered in the far west and the middle of the country; one off the south coast	Primarily 2,000–4,000 mm	Mostly crop and grazing land in the south and west, with forest in the north and east

167

Step 3: Investigate Central America prior to Hurricane Mitch

Q2. Fill in the Populated places Distribution column and the three Transportation columns in the table in Q1. Possible answers are listed in the table.

Q3. Analyze the annual precipitation for each country and fill in the Precipitation column in the table in Q1. Possible answers are listed in the table.

Q4. Fill in the last column in the table in Q1. Possible answers are listed in the table.

Q5. Which country has the largest proportion of its area devoted to crops? El Salvador

Q6. Which country is the most mountainous? Honduras

Q7. Which country has the largest proportion of its territory covered by roads? El Salvador

Q8. If you are stopping here, save the map document under a new name, record the new name and location, and exit ArcMap.
Document Example: ABC_Region7.mxd
Location Example: C:\Student\ABC

Step 4: Track Hurricane Mitch

Q9. When was Tropical Storm Mitch at this location? 10/22/21Z, or 9 PM on the twenty-second of October

Q10. What was Mitch's wind speed at this location? 40 mph

Q11. What are the latitude and longitude coordinates for Hurricane Mitch at this location? 14.3 latitude, –77.7 longitude

Q12. When was Hurricane Mitch at this location? 10/24/09Z, or 9 AM on the twenty-fourth of October

Q13. What was Mitch's wind speed at this location? 80 mph

Q14. When was Hurricane Mitch at this location? 10/27/21Z, or 9 PM on the twenty-seventh of October

Q15. What was Mitch's wind speed at this location? 135 mph

Q16. How long did it take for Tropical Storm Mitch to become a category 5 hurricane?
Hurricane – 5 time point: 10/26/12Z
Tropical_Storm time point: 10/24/03Z
Time difference: 00/02/09, or 2 days and 9 hours

Q17. What was Hurricane Mitch's maximum wind speed during this period? 155 mph

Step 5: Measure the size of the storm

Q18. What is the diameter of the eye of Hurricane Mitch? About 25 miles

Q19. Fill in the first row in the table on the following page. Approximate answers are listed in the table.

Image	Distance (miles)			Change from previous image
	Diameter of the eye	**Diameter of the storm**	**Between the eye and the coastline of Honduras**	
mitch2sat.tif	25	830	110	————————————
mitch3sat.tif	13	1000	50	Storm appears more intense and enlarged. It's closer to the coastline.
mitch4sat.tif	0 (not visible)	830	0 (on shore)	The eye is not visible, clouds are much thicker, but the spiral shape is still visible.
mitch5sat.tif	0 (not visible)	880	0 (on shore)	The cloud area is still large, but the spiral shape is gone.

Q20. Fill in the rest of the table above. Approximate answers are listed in the table.

Step 6: Analyze rainfall from Hurricane Mitch

Q21. What pattern do you notice in the amount of rainfall? The greatest amount of precipitation is on the southwest arm of the storm.

Q22. Is this a pattern you expected to find? Why or why not? Answers will vary.

Q23. What is the highest range of rainfall in the Rain4 layer? 24–29 inches

Q24. Which country received the majority of this heavy rain? Nicaragua

Q25. Describe the difference between the rainfall patterns on October 30 (Rain4 layer) and October 31 (Rain5 layer). The heaviest rainfall on October 30 was centered over the western coast of Nicaragua and southern El Salvador, with other rain bands extending due north and one off the eastern coast of Nicaragua and Honduras. On October 31, the main rain center was less intense. The outside bands appear to have merged with the main rain band from October 30.

Q26. What kind of damage do you expect to find with this type of storm? What aspects of the region will be most affected? Use the table in Q1 as a resource. Answers will vary and can mention flooding, landslides, damaged utility lines, washed out roads, destroyed crops, etc.

Q27. Record the new name of the document and its location.
Document Example: ABC_Region7.mxd
Location Example: C:\Student\ABC

Assessment rubrics

Middle school

Standard	Exemplary	Mastery	Introductory	Does not meet requirements
1 The student understands the characteristics, functions, and applications of maps, globes, aerial and other photographs, satellite-produced images, and models to solve geographic problems	Uses GIS to gather a variety of data about Hurricane Mitch including new data from outside sources; analyzes this information to identify cities at greatest risk for particular hazards	Uses GIS to gather a variety of data about Hurricane Mitch and analyzes this information to identify cities at greatest risk for particular hazards	Uses GIS to gather data about Hurricane Mitch and attempts to identify cities at risk for particular hazards; because of limited data, some predictions may not be accurate	Uses only data from the lesson and does not make accurate predictions
7 The student understands how to predict the consequences of physical processes on Earth's surface	Accurately predicts the impact of Hurricane Mitch on the physical environment of a country; creates a map of the affected areas using a variety of data	Accurately predicts the impact of Hurricane Mitch on the physical environment of a country; provides ample data to support predictions	Attempts to predict the impact of Hurricane Mitch on the physical environment of a country; provides some data to support predictions	Does not identify major effects of Hurricane Mitch on the physical environment of a country
15 The student understands how natural hazards affect human activities	Accurately predicts the impact of Hurricane Mitch on human activities in a country; creates a map of the affected areas using a variety of data	Accurately predicts the impact of Hurricane Mitch on human activities in a country; provides ample data to support predictions	Attempts to predict the impact of Hurricane Mitch on human activities in a country; provides some data to support predictions	Does not identify the effects of Hurricane Mitch on human activities in a country
18 The student understands how to apply the geographic point of view to solve social and environmental problems by making geographically informed decisions	Creates an emergency action plan for a country that takes into account infrastructure and environmental changes; uses a variety of data to support the plan; creates a map with evacuation routes and other important factors	Creates an emergency action plan for a country that takes into account infrastructure and environmental changes; uses a variety of data to support the plan, including an evacuation route	Creates an emergency action plan but takes into account only infrastructure or environmental changes; uses data to support the plan but misses some important factors	Creates an outline for an emergency action plan but takes into account only infrastructure or environmental changes; provides little or no data to support the plan

This is a four-point rubric based on the National Standards for Geographic Education. The mastery level meets the target objective for grades 5–8.

High school

	Standard	Exemplary	Mastery	Introductory	Does not meet requirements
1	The student understands how to use technologies to represent and interpret Earth's physical and human systems	Uses GIS to analyze a variety of data from satellite imagery to social infrastructure themes to determine what areas are at greatest risk from Hurricane Mitch; imports data from outside sources	Uses GIS to analyze a variety of data from satellite imagery to social infrastructure themes to determine what areas are at greatest risk from Hurricane Mitch	Uses GIS to analyze data to determine what areas are at greatest risk from Hurricane Mitch. Because of limited data, some predictions may not be accurate	Uses only data from the lesson and does not make accurate predictions
7	The student understands the spatial variation in the consequences of physical processes across Earth's surface	Accurately predicts and analyzes the impact of Hurricane Mitch on the physical environment of a country; creates a GIS map showing the storm's effects on the region	Accurately predicts and analyzes the impact of Hurricane Mitch on the physical environment of a country; provides details on the storm's effects on the region	Reviews data and attempts to make predictions about the storm's impact on the physical environment of a country	Makes inaccurate predictions on the storm's impact on the physical environment of a country
15	The student understands strategies of response to constraints placed on human systems by the physical environment and how humans perceive and react to natural hazards	Accurately predicts and analyzes the impact of Hurricane Mitch on human systems of a country; creates a GIS map detailing how the storm affects the region	Accurately predicts and analyzes the impact of Hurricane Mitch on human systems of a country; provides details on how the storm affects the region	Reviews data and attempts to make predictions about the storm's impact on human systems of a country	Makes inaccurate predictions on the storm's impact on human systems of a country
18	The student understands how to use geographic knowledge, skills, and perspectives to analyze problems and make decisions	Creates an emergency action plan for a country that takes into account infrastructure and environmental changes; uses a variety of data to support the plan; creates a map with evacuation routes and other important factors	Creates an emergency action plan for a country that takes into account infrastructure and environmental changes; uses a variety of data to support the plan, including an evacuation route	Creates an emergency action plan but takes into account only infrastructure or environmental changes; uses data to support the plan but misses some important factors	Creates an outline for an emergency action plan but takes into account only infrastructure or environmental changes; provides little or no data to support the plan

This is a four-point rubric based on the National Standards for Geographic Education. The mastery level meets the target objective for grades 9–12.

Module 7: Lesson 2

171

Bibliography

Center for International Earth Science Information Network (CIESIN),

The Earth Institute, Columbia University, NASA Socioeconomic Data and Applications Center (SEDAC). Elevation data [data file]. Available from Center for International Earth Science Information Network Web site, www.ciesin.columbia.edu.

Central Intelligence Agency. 2007. Economic data [data file]. Available from U.S.

Central Intelligence Agency World Factbook Web site, https://www.cia.gov/library/publications/ download/.

Chandler, T. 1989. *Four thousand years of urban growth: An historical census.* Lewiston, N.Y.: Edwin Mellen Press.

DeBlij, H. J., and P. O. Muller. 1992. *Geography: Regions and concepts.* New York: John Wiley & Sons, Inc.

ESRI. 2007. ESRI Data & Maps 2007 (update) [data DVD set]. Redlands, Calif.: ESRI.

———. 2006. ESRI Data & Maps 2006 [data DVD set]. Redlands, Calif.: ESRI.

———. 1999. *Getting to know ArcView GIS.* 3rd ed. Redlands, Calif.: ESRI Press.

———. 1996. ArcAtlas: Our earth [computer software]. Redlands, Calif.: ESRI.

Fellmann, J., A. Getis, and J. Getis. 1992. *Human geography: Landscapes of human activities.* 3rd ed. Dubuque, Iowa: Wm. C. Brown Publishers.

Ferrigno, J. G., and R. S. Williams. 1999. Satellite image atlas of glaciers of the world. U.S. Geological Survey Fact Sheet 130-02. Retrieved December 16, 2004, from pubs.usgs.gov/fs/fs 130-02/fs 130-02.html.

Food and Agricultural Organization of the United Nations. n.d. FAO Web site. Retrieved July 2001, from www.fao.org.

Geography Education Standards Project. 1994. *Geography for Life: National Geography Standards 1994.* Washington, D.C.: National Geographic Research and Exploration.

Guiney, J. L., and M. B. Lawrence. 1999, January 28. Preliminary report: Hurricane Mitch 22 October–05 November 1998. National Hurricane Center. Retrieved September 17, 2001, from www.nhc.noaa.gov/1998mitch.html.

Hardwick, S. W., and D. G. Holtgrieve. 1996. *Geography for educators: Standards, themes, and concepts.* Upper Saddle River, N.J.: Prentice Hall. 82–112.

Instituto Nacional de Estadistica, Geographica e Informatica, Mexico. 2001. Export and import data [data file]. Available from Instituto Nacional de Estadistica, Geographica e Informatica, Mexico, Web site, www.inegi.gob.mx.

International Society for Technology in Education. 1998. *National educational technology standards for all students.* Eugene, Ore.: Author.

Kious, J., and R. I. Tilling. 1996. *This dynamic earth: The story of plate tectonics.* Retrieved June 7, 2001, from pubs.usgs.gov/publications/text/dynamic.html.

McFalls, J. A., Jr. 1998. Population: A lively introduction. Population Reference Bureau Population Bulletin 53 (3).

National Aeronautics and Space Administration. n.d. NASA visible earth: Cryosphere. Retrieved December 16, 2004, from visibleearth.nasa.gov/Cryosphere/Sea_Ice/Icebergs.html.

National Aeronautics and Space Administration. n.d. NASA visible earth: Larsen B Ice Shelf breakup. Retrieved May 14, 2007, from http://svs.gsfc.nasa.gov/vis/a000000/a002400/a002421/index.html.

National Council for Teachers of Mathematics. 2000. *Principles and standards for school mathematics.* Reston, Va.: Author.

National Oceanic and Atmospheric Administration. 2001. The south geographic pole (Image ID: Corp1566, NOAA Corps Collection) [data file]. Available from NOAA Photo Library Web site, www.photolib.noaa.gov/corps/corp1566.htm.

National Research Council—National Academy of Sciences. 1995. *National science education standards.* Washington, D.C.: National Academy Press.

National Snow and Ice Data Center. n.d. Antarctic ice shelves and icebergs. Retrieved December 16, 2004, from nsidc.org/iceshelves.

NOVA. 1998, April. Warnings from the ice. Retrieved December 15, 2004, from www.pbs.org/wgbh/nova/warnings. WGBH Educational Foundation.

Ormsby, Tim, Eileen Napoleon, and Robert Burke. 2004. *Getting to know ArcGIS Desktop.* 2nd ed. Redlands, Calif.: ESRI Press.

Population Reference Bureau. n.d. PRB home page. Retrieved December 2004 from www.prb.org.

Schofield, Richard. Abridged version of speech given March 31, 1999, published in *British–Yemeni Society Journal* (July 2000) and retrieved from www.al-bab.com/bys/articles/schofield00.htm.

Sheets, B., and J. Williams. 2001. *Hurricane watch: Forecasting the deadliest storms on earth.* New York: Vintage Books.

Statistics Canada. 2007. Export and import data [data file]. Available from Statistics Canada Trade Calculator Web site, http://strategis.ic.gc.ca/sc_mrkti/tdst/tdo/tdo.php#tag.

United Nations. 2007. Estimates and projections of urban, rural and city populations. Available from United Nations Population Division Web site, http://esa.un.org/unup/index.asp?panel=1.

U.S. Census Bureau. 2007. Demographic data [data file]. Available from U.S. Census Bureau Web site, www.census.gov/ipc/www/.

———. 2007. Export and import data [data file]. Available from U.S. Census Bureau Foreign Trade Web site, www.census.gov/foreign-trade/www.

———. 2001. Export and import data [data file]. Available from U.S. Census Bureau Foreign Trade Web site, www.census.gov/foreign-trade/www.

———. 2007. World Vital Events Table [table image]. Available from U.S. Census Bureau Web site, www.census.gov/ipc/www/idb/worldpopinfo.html.

U.S. Energy Information Agency. 2007. Energy consumption data downloaded from the United States Energy Information Administration Web Site, www.eia.doe.gov/emeu/international/energyconsumption.html.

U.S. Geological Survey. 2001. Global GIS database: Digital atlas of Central and South America (Digital Data Series DDS-62-A) [computer software and data]. U.S.: Author.

———. 2001. Global GIS database: Digital atlas of the Middle East [computer software and data]. Unpublished, Author.

———. 2000, February. USGS TerraWeb: Antarctica. Retrieved December 16, 2004, from terraweb.wr.usgs.gov/projects/Antarctica.

———. n.d USGS National Earthquake Information Center. Retrieved April 14, 2007, from http://neic.usgs.gov/neis/epic/epic.html.

———. n.d. USGS Hurricane Mitch program. Retrieved September 10, 2001, from mitchnts1.cr.usgs.gov.

———. n.d. USGS volcano hazards program. Retrieved December 15, 2004, from volcanoes.usgs.gov.

Wade, T., and S. Sommer, eds. 2006. *A to Z GIS: An illustrated dictionary of geographic information systems.* Redlands, Calif. ESRI Press.

Whitaker, B. 2000, June 12. Translation of the Treaty of Jeddah. Retrieved from www.al-bab.com/yemen/pol/int5.htm.

———. 2000, July 1. Commentary on the border treaty. YEMEN Gateway. Retrieved from www.al-bab.com/yemen/pol/border000629.htm.

White, Frank. 1998, September. *Overview effect: Space exploration and human evolution.* 2nd ed. Reston, Va.: American Institute of Aeronautics and Astronautics.

Williams, J. 2001. Ice shelves float on the sea. USA Today.com. Retrieved September 6, 2001, from www.usatoday.com/weather/antarc/aiceshlf.htm.

———. 2001 Warming effect on sea level unsure. Edited summary of Intergovernmental Panel on Climate Change's book Climate Change 2001. USA Today.com. Retrieved September 6, 2001, from www.usatoday.com/weather/antarc/iceipcc.htm.

World Climate.com. n.d. Various world climate datasets [data file]. Available from World Climate.com Web site, www.worldclimate.com.

World geography: Building a global perspective. 1998. Upper Saddle River, N.J.: Prentice Hall.

World Wildlife Fund. n.d. Conservation science: Global 200 ecoregions. Retrieved December 15, 2004, from www.worldwildlife.org/science/ecoregions/g200.cfm.

ArcMap toolbar reference

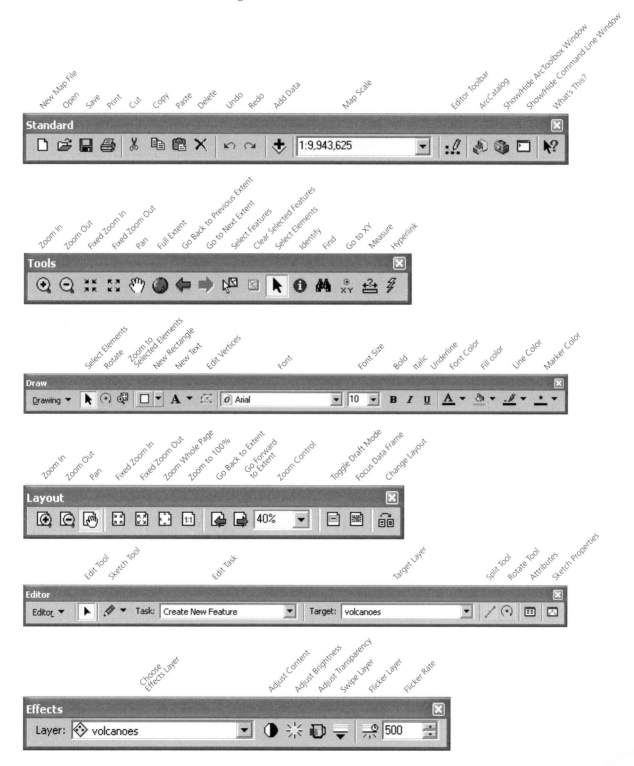

ArcMap zoom and pan tools

ArcMap gives you a number of ways to move around the map display. The better you know these tools, the more quickly you will be able to investigate different areas and features. Here are a few of them:

Tools for zooming in or out from the center of the map

Fixed Zoom In. Click the button to zoom in a fixed amount toward the center of the map. Click the button again to zoom in more.

Fixed Zoom Out. Click the button to zoom out a fixed amount away from the center of the map. Click the button again to zoom out more.

Tools for zooming in or out from anyplace on the map

Zoom In. Click the button, then click a spot on the map or drag a box around an area to zoom in on it. When the map redraws, the point or area you selected will appear in the center of the display.

Zoom Out. Click the button, then click a spot on the map or drag a box around an area to zoom out from it. When the map redraws, the point or area you selected will appear in the center of the display.

Tools for jumping to a specific map display

Full Extent. Click the Full Extent button to zoom to the entire map.

Go Back To Previous Extent. Click the button one or more times to go backward through earlier map displays you were browsing.

Go To Next Extent. Click the button one or more times to go forward again through the series of map displays you were browsing.

Tools for panning the map

Pan. Click the button. Click a spot on the map, hold down the mouse button, and drag it to a new location.

Scrollbars. Drag the scrollbars below and to the right of the map to pan the map side to side or up and down.

Zooming and panning with your mouse

If your mouse has a center wheel (a wheel between the left and right mouse buttons) you can use the wheel to zoom and pan the map display.

Zoom. Click a spot on the map. Roll the mouse wheel forward and backward to zoom in or out.

Pan. Move your mouse pointer over the map. Click with the mouse wheel and hold it down. Drag the mouse to pan the map.

Making quality maps

Think about the items below as you finish your maps or layouts. They will help you make sure your maps and layouts are the best they can be.

Map composition

Do your maps have the following elements?
- Title (addresses the major theme in your analysis)
- Legend
- Scale
- Compass rose
- Author (your name)
- Date map was created

Classification

Did you make reasonable choices for the classifications of the different layers on your maps? Is the symbology appropriate for the various layers?
- For quantitative data, is there a logical progression from low to high values and are they clearly labeled?
- For qualitative data, did you make sure not to imply any ranking in your legend?

Scale and projection

- Is the map scale appropriate for your problem?
- Have you used an appropriate map projection?

Implied analysis

- Did you correctly interpret the color, pattern, and shape of your symbologies?
- Have you included text to inform the reader of the map's intended use?

Design and aesthetics

- Are your maps visually balanced and attractive?
- Can you distinguish the various symbols for different layers in your maps?

Effectiveness of map

- How well do the map components communicate the story of your map?
- Do the map components take into account the interests and expertise of the intended audience?
- Are the map components of appropriate size?

181

GIS terms

ArcGIS

Computer software for implementing a geographic information system (GIS).

ArcView

Desktop GIS software that includes ArcMap for displaying and interacting with maps and layouts and ArcCatalog for previewing data and metadata.

attribute

A piece of information that describes a geographic feature on a GIS map. The attributes of an earthquake might include the date it occurred and its latitude, longitude, depth, and magnitude.

attribute table

A table that contains all of the attributes for like features on a GIS map, arranged so that each row represents one feature and each column represents one feature attribute. In a GIS, attribute values in an attribute table can be used to find, query, and symbolize features.

OBJECTID *	Shape *	CITY_NAME	CNTRY_NAME	ANCIENT	RANK_1950	POP_1950
1	Point	Moscow	Russia	Moscow	6	5,100,000
2	Point	London	United Kingdom	London	2	8,860,000
3	Point	Paris	France	Paris	4	5,900,000
4	Point	Chicago	United States	Chicago	8	4,906,000
5	Point	New York	United States	New York	1	12,463,000
6	Point	Tokyo	Japan	Edo	3	7,000,000
7	Point	Shanghai	China	Shanghai	5	5,406,000
8	Point	Calcutta	India	Calcutta	10	4,800,000
9	Point	Buenos Aires	Argentina	Buenos Aires	7	5,000,000
10	Point	Essen/Ruhr area	Germany	<Null>	9	4,900,000

Attributes of Top 10 Cities, 1950 C.E.

The attribute table for the Top 10 Cities, 1950 C.E. layer includes attributes for each of the ten cities listed.

axis

The vertical (y-axis) or horizontal (x-axis) lines in a graph on which measurements can be illustrated and coordinated with each other. Each axis in a GIS graph can be made visible or invisible and labeled.

bookmark

In ArcMap, a shortcut you can create to save a particular geographic extent on a map so you can return to it later. Also known as a spatial bookmark.

color selector

The window that allows you to change the color of geographic features and text on your GIS map.

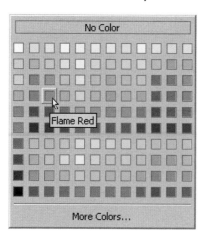

comma-delimited values file (.csv)

A data table in text form where the values are separated by commas. This is a popular format for transferring data from one program to another, for example between spreadsheet programs and ArcMap. These programs use the commas to determine where a new piece of data starts and stops.

coordinate system

A system of intersecting lines that is used to locate features on surfaces such as the earth's surface or a map. In ArcMap, each feature class (layer) of data has a coordinate system that tells ArcMap where on the map to draw the features. A feature class may also have a map projection. See also feature class and map projection.

data

Any collection of related facts, from raw numbers and measurements to analyzed and organized sets of information.

data folder

A folder on the hard drive of your computer or your network's computer that is available for storage of GIS data and map documents that you create.

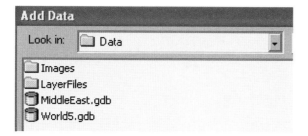

data frame

A map element that defines a geographic extent, a page extent, a coordinate system, and other display properties for one or more layers in ArcMap. In data view, only one data frame is displayed at a time; in layout view, all of a map's data frames are displayed at the same time.

data frame, active

In ArcMap, the active data frame is the target for many tools and commands. In data view, the active data frame is bold in the table of contents and visible in the display area. In layout view, the active data frame has a dashed line around it to show it is the active one.

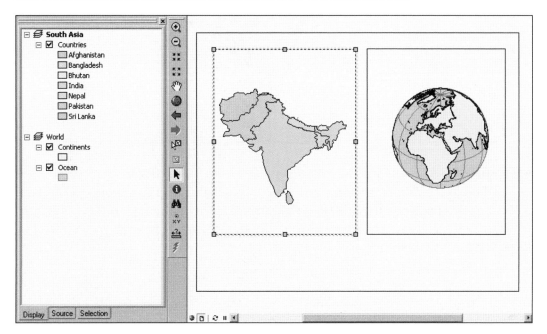

This map document, shown in layout view, has two data frames: South Asia and The World. South Asia is the active data frame.

data source

The data referenced by a layer or a layer file in ArcMap or ArcCatalog. Examples of data sources are a geodatabase feature class, a shapefile, and an image.

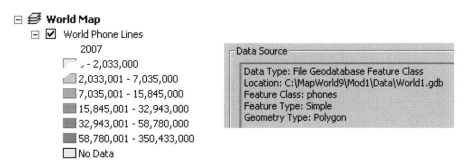

The data source for this World Phone Lines layer is the geodatabase feature class "phones" found in the World1 geodatabase (World1.mdb). The geodatabase is located in the C:\MapWorld\Mod1\ Data folder.

data view

A view in ArcMap for exploring, displaying, and querying geographic data. This view hides map elements such as titles, north arrows, and scale bars. Compare layout view.

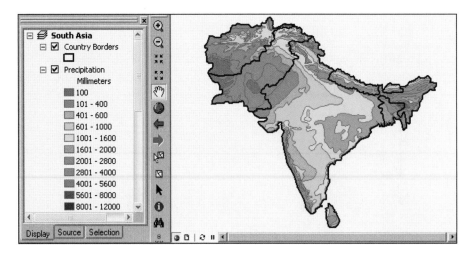

decimal degrees

Degrees of latitude and longitude expressed in decimals instead of minutes and seconds. Minutes and seconds are converted into a decimal using the mathematical formula below. In a GIS, using decimal degrees is more efficient than using minutes and seconds because it makes digital storage of coordinates easier and computations faster.

Decimal degrees = degrees + (minutes/60) + (seconds/3,600)

73° 59' 15" longitude = 73.9875 decimal degrees

feature

A geographic object on a map represented by a point, a line, or a polygon.

- A point feature is a map object that has no length or width, such as a tree on a neighborhood map or a city on a world map.
- A line feature is a one-dimensional map object such as a river or a street.
- A polygon feature is a two-dimensional map object such as a lake, a city, or a continent.

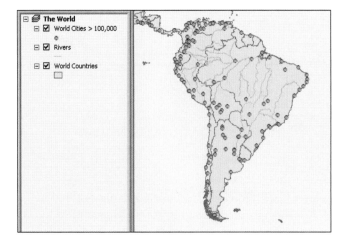

feature class

A collection of geographic features with the same geometry type (point, line, or polygon), the same attributes, and the same spatial reference (coordinate system and map projection).

field

The column in a table that contains the values (information) for a single attribute of each geographic feature in a GIS layer.

ObjectID	Shape *	City Name	Country Name	Population	CAPITAL
0	Point	Guatemala	Guatemala	1,400,000	Y
1	Point	Tegucigalpa	Honduras	551,606	Y
2	Point	San Salvador	El Salvador	920,000	Y
3	Point	Managua	Nicaragua	682,000	Y
4	Point	San Jose	Costa Rica	670,000	Y
5	Point	Belmopan	Belize	4,500	Y
6	Point	Panama	Panama	625,000	Y
7	Point	San Jose	US	629,400	N

In this table, the City Name field contains the name for each city in this layer. The Population field contains the population value for each city.

field name

The column heading in an attribute table. Because field names are often abbreviated, ArcGIS allows you to create an alternative name, or alias, that can be more descriptive. In the graphic above, City Name and Country Name are aliases for fields named "NAME" and "COUNTRY."

Find button

An ArcMap button used for locating one or more map features that have a particular attribute value.

folder connection

A shortcut that allows you to navigate to a folder without having to enter the entire path.

geodatabase

A database used to organize and store geographic data in ArcGIS.

georeference

To assign coordinates from a reference system, such as latitude/longitude, to the page coordinates of an image or map.

graduated color map

A map that uses a range of colors to show a sequence of numeric values. For example, on a population density map, the more people per square kilometer, the darker the color.

graph

A graphic representation of tabular data.

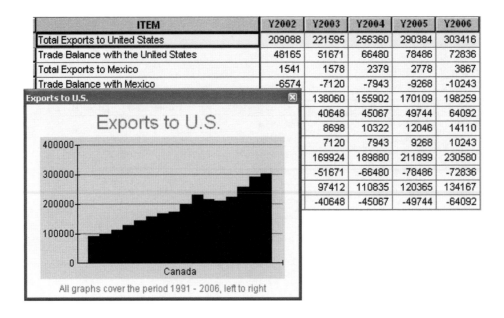

ITEM	Y2002	Y2003	Y2004	Y2005	Y2006
Total Exports to United States	209088	221595	256360	290384	303416
Trade Balance with the United States	48165	51671	66480	78486	72836
Total Exports to Mexico	1541	1578	2379	2778	3867
Trade Balance with Mexico	-6574	-7120	-7943	-9268	-10243
		138060	155902	170109	198259
		40648	45067	49744	64092
		8698	10322	12046	14110
		7120	7943	9268	10243
		169924	189880	211899	230580
		-51671	-66480	-78486	-72836
		97412	110835	120365	134167
		-40648	-45067	-49744	-64092

Identify tool

An ArcMap tool used to display the attributes of features in the map.

image

A graphic representation of data such as a scanned picture or a satellite photograph.

join

An operation that appends the fields of one table to those of another through an attribute field common to both tables. A join is usually used to attach more attributes to the attribute table of a map layer so that these attributes can be mapped. For example, you could join a country table with population data to a country layer attribute table. Compare relate.

label

Text placed next to a geographic feature on a map to describe or identify it. Feature labels usually come from an attribute field in the attribute table.

layer

A set of geographic features of the same type along with the associated attribute table, or an image. Examples of layers are "Major Cities," "Countries," and "Satellite Image." A layer references a specific data source such as a geodatabase feature class or image. Layers have properties, such as a layer name, symbology, and label placement. They can be stored in map documents (.mxd) or saved individually as layer files (.lyr). See also data source.

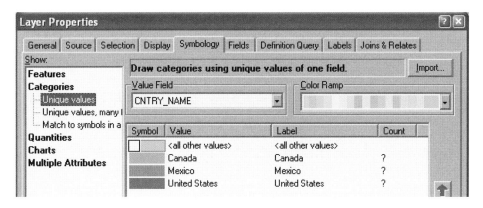

A layer has many properties, including Symbology properties. Some of the properties for the NAFTA Countries layer are pictured here.

layer, turn on

Turning on a layer allows the layer to be displayed in the map. In ArcMap, a layer is turned on by placing a check mark in the box next to the layer name in the table of contents.

layer file

In ArcGIS, a file with a .lyr extension that stores the path to a data source and other layer properties, including symbology.

layout

In ArcMap, an on-screen presentation document that can include maps, graphs, tables, text, and images.

layout view

A view in ArcMap in which geographic data and map elements, such as titles, legends, and scale bars, are placed and arranged for printing.

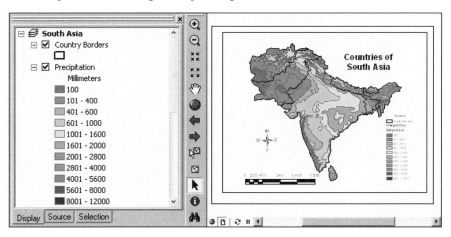

legend

A list identifying what each symbol represents on the map.

line

See feature.

Magnifier window

A window in ArcMap data view that shows a zoomed-in view of a small area of the main map. Moving the Magnifier window around does not change the extent of the map underneath.

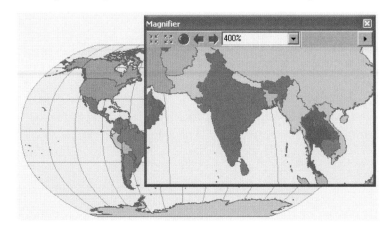

map document

In ArcMap, the file that contains one or more data frames and the associated layers, tables, graphs, and reports. Map document files have a .mxd extension.

map projection

A method by which the curved surface of the earth is portrayed on a flat map. Every map projection distorts distance, area, shape, direction, or some combination thereof. Map projections are made using complex mathematical formulas that are part of ArcGIS software's automatic functions.

MapTip

In ArcMap, a pop-up label for a map feature that is displayed when the mouse is paused over that feature. The label comes from a field in the layer attribute table.

Measure tool

An ArcMap tool used to measure distance on a map.

metadata

Information about the content, quality, condition, and other characteristics of data. Metadata may include a brief description of the data and its purpose, the names of the authors or compilers of the data, the date it was collected or created, the meaning of attribute fields, its scale, and its spatial reference (coordinate system and map projection).

pan

To move your map up, down, or sideways without changing the viewing scale.

point

See feature.

polygon

See feature.

projection

See map projection.

record

A row in an attribute table that contains all of the attribute values for a single feature.

	ObjectID	Shape *	City Name	Country Name	Population
	0	Point	Guatemala	Guatemala	1,400,000
	1	Point	Tegucigalpa	Honduras	551,606
	2	Point	San Salvador	El Salvador	920,000
▶	3	Point	Managua	Nicaragua	682,000
	4	Point	San Jose	Costa Rica	670,000
	5	Point	Belmopan	Belize	4,500
	6	Point	Panama	Panama	625,000
	7	Point	San Jose	US	629,400

This table has seven records. The fourth record is highlighted. It contains all of the attributes for the point feature representing the city of Managua, Nicaragua.

relate

An operation that establishes a temporary connection between records in two tables using a field common to both. Unlike a join operation, a relate does not append the fields of one table to the other. A relate is usually used to associate more records and their attributes to the attribute table of a map layer. For example, you could relate a table listing large cities to a layer attribute table of countries, or you could join a world cities table to a country layer attribute table. Compare join.

scale

The relationship between a distance or area on a map and the corresponding distance or area on the ground, commonly expressed as a fraction or ratio. A map scale of 1/100,000 or 1:100,000 means that one unit of measure (e.g., one inch) on the map equals 100,000 of the same units on the earth.

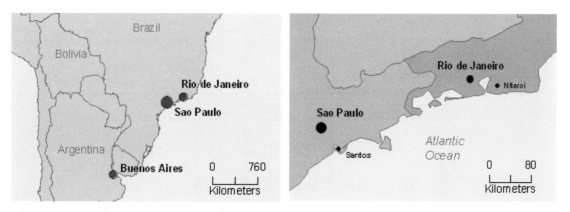

The map on the left has a map scale of 1:80,000,000. The map on the right, which is zoomed in, has a map scale of 1:6,000,000.

selected feature

A geographic feature that is chosen and put into a subset so that various functions can be performed on the feature. In ArcMap, a feature can be selected by clicking it on the map with the Select Features tool or highlighting an attribute in a table. When a geographic feature is selected, it is outlined in blue on the map. Its corresponding record in the attribute table is highlighted in blue.

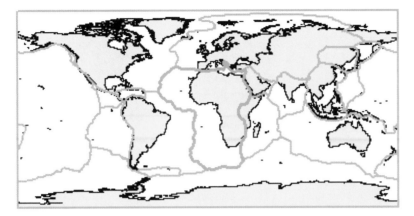

The Africa plate is selected in this map of the earth's tectonic plates.

shapefile (.shp)

A data storage format for storing the location, shape, and attribute information of geographic features. A shapefile is stored in a set of related files and contains one feature class.

source data

See data source.

sort ascending

To arrange an attribute table's rows in order from the lowest values to the highest values in a field. For example, number values would be ordered from 1 to 100, and alphabetical values would be ordered from A to Z.

sort descending

To arrange an attribute table's rows in order from the highest to the lowest values in a field. For example, number values would be ordered from 100 to 1, and alphabetical values would be ordered from Z to A.

Symbol Selector

The dialog in ArcMap for selecting symbols and changing their color, size, outline, or other properties.

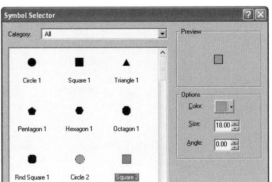

table of contents

A list of data frames and layers on a map that may also show how the data is symbolized.

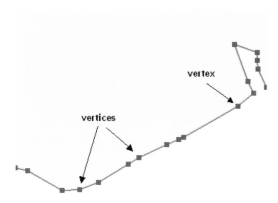

```
☐ ⧉ The World
   ☐ ☑ World Cities > 100,000
            ●
   ☐ ☑ Rivers

   ☐ ☑ World Countries
            ☐

☐ ⧉ Standard of Living Indicators
   ⊞ ☐ Infant Mortality Rate
   ☐ ☑ Life Expectancy
            Years
            ☐ 32 - 46
            ☐ 47 - 57
            ☐ 58 - 67
            ▨ 68- 75
            ▦ 76 - 84
            ☐ No Data
   ⊞ ☐ Literacy Rate
```

toolbar

A set of commands that allows you to carry out related tasks. The Main Menu toolbar in ArcMap has a set of menu commands; other toolbars typically have buttons. Toolbars can float on the desktop in their own window or may be docked at the top, bottom, or sides of the main window.

vertex

One of the points that defines a line or polygon feature.

zoom

To display a larger or smaller extent of a GIS map or image.

References and image credits

References

Geography Education Standards from Geography Education Standards Project. 1994. *Geography for Life: National Geography Standards 1994.* Washington, D.C.: National Geographic Research and Exploration.

Image Credits

Introduction

Page xii: Courtesy of Photodisc.

Page 4: Maps by ESRI (for data layer information see "Data Sources").

Page 8: Map and illustration by ESRI (for data layer information see "Data Sources").

Page 9: Map and illustration by ESRI (for data layer information see "Data Sources").

Module 1

Page 12: Photo by Janis Christie, courtesy of Photodisc/Getty Images.

Page 14: Map by ESRI (for data layer information see "Data Sources").

Page 22: Map by ESRI (for data layer information see "Data Sources").

Module 2

Page 32: Photo by InterNetwork Media, courtesy of Photodisc/Getty Images.

Page 34: Map by ESRI (for data layer information see "Data Sources").

Page 44: Map by ESRI (for data layer information see "Data Sources").

Module 3

Page 52: Photo by D. Normark/PhotoLink, courtesy of Photodisc/Getty Images.

Page 54: Map by ESRI (for data layer information see "Data Sources").

Page 68: Map by ESRI (for data layer information see "Data Sources").

Module 4

Page 80: Courtesy of Photodisc/Getty Images.

Page 82: Map by ESRI (for data layer information see "Data Sources").

Page 94: Map by ESRI (for data layer information see "Data Sources").

Module 5

Page 105: Photo by The Studio Dog, courtesy of Photodisc/Getty Images.

Page 106: Map by ESRI (for data layer information see "Data Sources").

Page 116: Map by ESRI (for data layer information see "Data Sources").

Module 6

Page 128: Photo by Stockbyte, courtesy of Photodisc/Getty Images.

Page 131: Map by ESRI (for data layer information see "Data Sources").

Page 140: Map by ESRI (for data layer information see "Data Sources").

Module 7

Page 150: Photo by PhotoLink, courtesy of Photodisc/Getty Images.

Page 152: Map by ESRI (for data layer information see "Data Sources").

Page 162: Map by ESRI (for data layer information see "Data Sources").

Data sources

Module 1 - \OurWorld2\Mod1\Data
Data sources include:
\World1.gdb\Cities_GR_100K feature class from ESRI Data & Maps 2006, courtesy of ArcWorld.

\World1.gdb\Cities_GR_5M feature class from ESRI Data & Maps 2006, courtesy of ArcWorld.

\World1.gdb\cntry07_demog combined feature class from ESRI Data & Maps 2007, courtesy of ArcWorld; the US Census Bureau, International Division; and the CIA World Factbook.

\World1.gdb\cntry07_econ combined feature class from ESRI Data & Maps 2007, courtesy of ArcWorld and the CIA World Factbook.

\World1.gdb\geogrid feature class from ESRI Data & Maps 2006.

\World1.gdb\lakes feature class from ESRI Data & Maps 2006, courtesy of ArcWorld.

\World1.gdb\rivers feature class from ESRI Data & Maps 2006, courtesy of ArcWorld.

Module 2 - \OurWorld2\Mod2\Data
Data sources include:
\Images\earth_wsi.sid from ESRI Data & Maps 2004, satellite image courtesy of WorldSat, Inc.

\EastAsia.gdb\popdens05 feature class from Center for International Earth Science Information Network (CIESIN), The Earth Institute, Columbia UniversityArchives Manager, and NASA Socioeconomic Data and Applications Center (SEDAC).

\World2.gdb\cntry07 feature class from ESRI Data & Maps 2007, courtesy of ArcWorld

\World2.gdb\continents feature class from ESRI Data & Maps 2004, courtesy of ArcWorld Supplement.

\World2.gdb\earthquakes feature class from U.S. Geographical Survey National Earthquake Information Center.

\World2.gdb\faults feature class from ArcAtlas: Our Earth, 1996, courtesy of Data+ and ESRI.

\World2.gdb\features feature class from Mapping Our World: GIS Lessons for Educators ArcGIS Desktop Edition, ESRI Press, 2005. Original data courtesy of National Geographic Maps.

\World2.gdb\major_cities feature class from ESRI Data & Map 2004, courtesy of DMTI Spatial Inc.

\World2.gdb\plates feature class from ArcAtlas: Our Earth, 1996, courtesy Data+ and ESRI, and edited by Roger Palmer based on ocean bathymetry from WSI_Earth.sid in comparison with USGS plate lines.

\World2.gdb\plates_line feature class from ArcAtlas: Our Earth, 1996, courtesy Data+ and ESRI, and edited by Roger Palmer based on ocean bathymetry from WSI_Earth.sid in comparison with USGS plate lines.

\World2.gdb\volcanoes feature class from ArcAtlas: Our Earth, 1996, courtesy Data+ and ESRI; created from plate lines data edited by Roger Palmer based on ocean bathymetry from WSI_Earth. sid in comparison with USGS plate lines.

\World2.gdb\world30 feature class from ESRI Data & Maps 2004.

Module 3 - \OurWorld2\Mod3\Data
Data sources include:

\SouthAsia.gdb\agriculture feature class from *Mapping Our World: GIS Lessons for Educators ArcGIS Desktop Edition,* ESRI Press, 2005; original data courtesy of Data+ and ESRI, ArcAtlas: Our Earth.

\SouthAsia.gdb\cities_sa feature class from ESRI Data & Maps, 2004.

\SouthAsia.gdb\cntry_sa feature class from *Mapping Our World: GIS Lessons for Educators ArcGIS Desktop Edition,* ESRI Press, 2005; original data from ESRI Data & Maps 2001, courtesy of ArcWorld.

\SouthAsia.gdb\pop_dens feature class from *Mapping Our World: GIS Lessons for Educators ArcGIS Desktop Edition,* ESRI Press, 2005; original data courtesy of Data+ and ESRI, ArcAtlas: Our Earth.

\SouthAsia.gdb\precip_sa feature class from *Mapping Our World: GIS Lessons for Educators ArcGIS Desktop Edition,* ESRI Press, 2005; original data courtesy of Data+ and ESRI, ArcAtlas: Our Earth.

\SouthAsia.gdb\relief_sa feature class from *Mapping Our World: GIS Lessons for Educators ArcGIS Desktop Edition,* ESRI Press, 2005; original data courtesy of Data+ and ESRI, ArcAtlas: Our Earth.

\SouthAsia.gdb\rivers_sa feature class from *Mapping Our World: GIS Lessons for Educators ArcGIS Desktop Edition,* ESRI Press, 2005; original data courtesy of Data+ and ESRI, ArcAtlas: Our Earth.

\World3.gdb\climate feature class from *Mapping Our World: GIS Lessons for Educators ArcGIS Desktop Edition,* ESRI Press 2005; original data courtesy of National Geographic Maps.

\World3.gdb\cntry07_demog feature class from ESRI Data & Maps 2007 Update, courtesy of ArcWorld Supplement; US Census Bureau, International Division; and the CIA World Factbook.

\World3.gdb\geogrid feature class from ESRI Data & Maps 2004.

\World3.gdb\lakes feature class from ESRI Data & Maps 2004, courtesy of ArcWorld.

\World3.gdb\rivers feature class from ESRI Data & Maps 2004, courtesy of ArcWorld.

\World3.gdb\temp_city combined feature class from Worldclimate.com and ESRI Data and Maps 2004, author added Celsius in addition to Fahrenheit to the data.

\World3.gdb\world30 feature class from ESRI Data & Maps 2004.

\WorldPopDensity05.gdb\WorldPopDensity05 feature class from Center for International Earth Science Information Network (CIESIN) the Earth Institute, Columbia University

Archives Manager, and NASA Socioeconomic Data and Applications Center (SEDAC).

Module 4 - \OurWorld2\Mod4\Data
Data sources include:

\World4.gdb\TopTen\city100 feature class derived from Tertius Chandler's book, *Four Thousand Years of Urban Growth: An Historical Census,* Edwin Mellen Press,1987.

\World4.gdb\TopTen\city1000 feature class derived from Tertius Chandler's book, *Four Thousand Years of Urban Growth: An Historical Census,* Edwin Mellen Press,1987.

\World4.gdb\TopTen\city1500 feature class derived from Tertius Chandler's book, *Four Thousand Years of Urban Growth: An Historical Census,* Edwin Mellen Press,1987.

\World4.gdb\TopTen\city1800 feature class derived from Tertius Chandler's book, *Four Thousand Years of Urban Growth: An Historical Census,* Edwin Mellen Press,1987.

\World4.gdb\TopTen\city1900 feature class derived from Tertius Chandler's book, *Four Thousand Years of Urban Growth: An Historical Census,* Edwin Mellen Press,1987.

\World4.gdb\TopTen\city1950 feature class derived from Tertius Chandler's book, *Four Thousand Years of Urban Growth: An Historical Census,* Edwin Mellen Press,1987.

\World4.gdb\TopTen\city2000 combined feature class from ESRI Data & Maps 2006 (cities) and the United Nations Population Division (population).

\World4.gdb\TopTen\city2005 combined feature class from ESRI Data & Maps 2006 (cities) and the United Nations Population Division (population).

\\World4.gdb\cntry07_econ combined feature class from ESRI Data & Maps 2007 courtesy of ArcWorld and the CIA World Factbook.

\World4.gdb\cntry07_social combined feature class from ESRI Data & Maps 2007, courtesy of ArcWorld; the U.S. Census Bureau, International Division; and the CIA World Factbook.

\World4.gdb\continents feature class from ESRI Data & Maps 2004, courtesy of ArcWorld Supplement.

\World4.gdb\geogrid feature class from ESRI Data & Maps 2004.

\World4.gdb\lakes feature class from ESRI Data & Maps 2004, courtesy of ArcWorld.

\World4.gdb\rivers feature class from ESRI Data & Maps 2004, courtesy of ArcWorld.

\World4.gdb\top30cities combined feature class from ESRI Data & Maps 2006 (cities) and United Nations Population Division (population).

\World4.gdb\vital_statistics combined feature class from ESRI Data & Maps 2007, courtesy of ArcWorld; the U.S. Census Bureau, International Division; and the CIA World Factbook.

\World4.gdb\world30 feature class from ESRI Data & Maps 2004.

Module 5 - \OurWorld2\Mod5\Data
Data sources include:

\Images\earth_wsi.sid courtesy of WorldSat International, Inc.

\Images\middle_east.tif courtesy of USGS, 2001, Global GIS Database: Digital Atlas of Africa.

\MiddleEast.gdb\Boundaries\ap_cntry feature class from ESRI Data & Maps 2000, courtesy of ArcWorld Supplement.

\MiddleEast.gdb\ Boundaries\ap_line feature class from ESRI Data & Maps 2004, courtesy of ArcWorld Supplement.

\MiddleEast.gdb\ Boundaries\arab_pen feature class from ESRI Data & Maps 2004, courtesy of ArcWorld Supplement.

\MiddleEast.gdb\ Boundaries\ Arabian_Peninsula___namesAnno feature class derived from ESRI Data & Maps, 2004, courtesy of ArcWorld Supplement.

\MiddleEast.gdb\ Boundaries\neighbors feature class from ESRI Data & Maps 2004, courtesy of ArcWorld Supplement.

\MiddleEast.gdb\ Boundaries\yemen1 feature class digitized from paper maps: *British-Yemeni Society Journal,* 2000.

\MiddleEast.gdb\ Boundaries\yemen2 feature class digitized from paper maps: *British-Yemeni Society Journal,* 2000.

\MiddleEast.gdb\ Boundaries\yemen3 feature class digitized from paper maps: *British-Yemeni Society Journal,* 2000.

\MiddleEast.gdb\Cities\city_town feature class from USGS; Digital Atlas of the Middle East (NIMA data).

\MiddleEast.gdb\Cities\major_cities feature class from USGS, 2001, Global GIS Database: Digital Atlas of Africa; from file gaz_cities.shp

\MiddleEast.gdb\agriculture feature class from ArcAtlas: Our Earth; Data+ and ESRI, 1996.

\MiddleEast.gdb\eco_zone feature class from ESRI Data & Maps 2004, courtesy of World Wildlife Fund.

\MiddleEast.gdb\pop_density feature class from ArcAtlas: Our Earth, courtesy Data+ and ESRI, 1996.

\MiddleEast.gdb\precipitation feature class from USGS, 2001, Global GIS Database: Digital Atlas of Africa (NIMA data).

\MiddleEast.gdb\roads feature class from USGS, 2001, Global GIS Database: Digital Atlas of Africa (NIMA data).

\MiddleEast.gdb\springs feature class from USGS, 2001, Global GIS Database: Digital Atlas of Africa (NIMA data).

\MiddleEast.gdb\streams feature class from USGS, 2001, Global GIS Database: Digital Atlas of Africa (NIMA data).

\MiddleEast.gdb\temperature feature class from USGS, 2001, Global GIS Database: Digital Atlas of Africa (NIMA data).

\MiddleEast.gdb\waterbodies feature class from USGS, 2001, Global GIS Database: Digital Atlas of Africa (NIMA data).

\World5.gdb\cntry07_demog combined feature class from ESRI Data & Maps 2007, courtesy of ArcWorld; the US Census Bureau, International Division; and the CIA World Factbook.

\World5.gdb\cntry92 feature class from ESRI Data & Maps 2004, courtesy of ArcWorld.

\World5.gdb\lakes feature class from ESRI Data & Maps, courtesy of ArcWorld.

\World5.gdb\Language feature class from Mapping Our World: GIS Lessons for Educators ArcGIS Desktop Edition, ESRI Press, 2005; original data courtesy of National Geographic Maps.

\World5.gdb\oil_gas feature class from ArcAtlas: Our Earth; Data+ and ESRI, 1996.

\World5.gdb\Religion ESRI feature class from Mapping Our World: GIS Lessons for Educators ArcGIS Desktop Edition, ESRI Press, 2005; original data courtesy of National Geographic Maps.

\World5.gdb\rivers feature class from ESRI Data & Maps 2004, courtesy of ArcWorld.

\World5.gdb\world30 feature class from ESRI Data & Maps 2004.

Module 6 - \OurWorld2\Mod6\Data\
Data sources include:

\NAFTA.gdb\NAFTA_countries combined feature class from ESRI Data & Maps 2004 and U.S. Census.

\NAFTA.gdb\NAFTA_Trading_Statistics combined feature class from ESRI Data & Maps 2004 and U.S. Census.

\World6.gdb\cntry04_energy combined feature class from ESRI Data & Maps 2004 and the United States Energy Information Administration.

\World6.gdb\cntry04_social combined feature class from ESRI Data & Maps 2007 and the CIA World Factbook.

\World6.gdb\cntry07_demog combined feature class from ESRI Data & Maps 2007, courtesy of ArcWorld; the U.S. Census Bureau, International Division; and the CIA World Factbook.

\World6.gdb\cntry07_econ combined feature class from ESRI Data & Maps 2007, courtesy of ArcWorld and the CIA World Factbook.

Module 7 - \OurWorld2\Mod7\Data
Data sources incude:

\Images\amery.tif created by USGS using Digital Elevation Model data from USGS TerraWeb; http://TerraWeb.wr.usgs.gov 2001.

\Images\ellswrth.tif created by USGS using Digital Elevation Model data from USGS TerraWeb; http://TerraWeb.wr.usgs.gov 2001.

\Images\larsen_breakup.tif recorded by NASA's MODIS satellite sensor: NASA/Goddard Space Flight Center.

\Images\mcmurdo.tif created by USGS using Digital Elevation Model data from USGS TerraWeb; http://TerraWeb.wr.usgs.gov 2001.

\Images\min120m.jpg created using GTOPO30 and Spatial Analyst 2001.

\Images\mitch2sat.tif from USGS Digital Atlas of Central America, NOAA, NASA.

\Images\mitch3sat.tif from USGS Digital Atlas of Central America, NOAA, NASA.

\Images\mitch4sat.tif from USGS Digital Atlas of Central America, NOAA, NASA.

\Images\mitch5sat.tif from USGS Digital Atlas of Central America, NOAA, NASA.

\Images\plus50m.jpg created using GTOPO30 and Spatial Analyst 2002.

\Images\plus5m.jpg created using GTOPO30 and Spatial Analyst 2002.

\Images\plus73m.jpg created using GTOPO30 and Spatial Analyst 2002.

\Images\sealevel.jpg created using GTOPO30 and Spatial Analyst 2002.

\Images\southpole2.tif from NOAA Corps Collection.

\Antarctica.gdb\S_Pole_pts feature class digitized by ESRI from paper map, "Political and Physical Map of the World," courtesy National Geographic Society.

\CentralAmerica.gdb\agr_use feature class from ArcAtlas Our Earth, courtesy of Data+ and ESRI.

\CentralAmerica.gdb\ca_airportsfeature class from U.S. Geological Survey, Global GIS Database: Digital Atlas of Central America.

\CentralAmerica.gdb\ca_capitals feature class from ESRI Data & Maps 2000, courtesy of ArcWorld.

\CentralAmerica.gdb\ca_countries feature class from ESRI Data & Maps 2000, courtesy of ArcWorld Supplement.

\CentralAmerica.gdb\ca_drain data origin Digital Chart of the World gazetteer.

\CentralAmerica.gdb\ca_railroads feature class from U.S. Geological Survey, Global GIS Database: Digital Atlas of Central America.

\CentralAmerica.gdb\ca_roads feature class from U.S. Geological Survey, Global GIS Database: Digital Atlas of Central America.

\CentralAmerica.gdb\ca_utility feature class from U.S. Geological Survey, Global GIS Database: Digital Atlas of Central America.

\CentralAmerica.gdb\coastal data origin Digital Chart of the World gazetteer.

\CentralAmerica.gdb\landforms data origin Digital Chart of the World gazetteer.

\CentralAmerica.gdb\mitch2 courtesy of U.S. Geological Survey Digital Atlas of Central America, NOAA.

\CentralAmerica.gdb\mitch3 courtesy of U.S. Geological Survey Digital Atlas of Central America, NOAA, NASA.

\CentralAmerica.gdb\mitch4 courtesy of U.S. Geological Survey Digital Atlas of Central America, NOAA, NASA.

\CentralAmerica.gdb\mitch5 courtesy of U.S. Geological Survey Digital Atlas of Central America, NOAA, and NASA.

\CentralAmerica.gdb\mitch_pre feature class from U.S. Geological Survey, Digital Atlas of Central America & NOAA.

\CentralAmerica.gdb\pop_plc feature class from U.S. Geolgocial Survey, Digital Atlas of Central America & NOAA.

\CentralAmerica.gdb\precip feature class from ArcAtlas: Our Earth, 1996. courtesy of Data+ and ESRI.

\CentralAmerica.gdb\rain3 feature class from USGS Digital Atlas of Central America & NOAA.

\CentralAmerica.gdb\rain4 feature class from USGS Digital Atlas of Central America & NOAA.

\CentralAmerica.gdb\rain5 feature class from USGS Digital Atlas of Central America & NOAA.

\World7.gdb\airports feature class from from ArcAtlas: Our Earth, 1996, courtesy of Data+ and ESRI.

\World7.gdb\cntry07_demog combined feature class from ESRI Data & Maps 2007, courtesy of ArcWorld and the CIA World Factbook.

\World7.gdb\continents feature class from ESRI Data & Maps 2004, courtesy of ArcWorld Supplement.

\World7.gdb\energy feature class from ArcAtlas: Our Earth, 1996, courtesy of Data+ and ESRI.

\World7.gdb\geogrid feature class from ESRI Data & Maps 2006.

\World7.gdb\lakes feature calss from ESRI Data & Maps 2006, courtesy of ArcWorld.

\World7.gdb\latlong feature class from ESRI Data & Maps 2004.

\World7.gdb\major_cities feature class from ESRI Data & Maps 2000, courtesy of ArcWorld.

\World7.gdb\manufact_plc feature class from ArcAtlas: Our Earth, 1996, courtesy of Data+ and ESRI.

\World7.gdb\mineral_res feature class from ArcAtlas: Our Earth, 1996, courtesy of Data+ and ESRI.

\World7.gdb\pipelines feature class from ArcAtlas: Our Earth, 1996, courtesy of Data+ and ESRI.

\World7.gdb\rivers feature class from ESRI Data & Maps 2006, courtesy of ArcWorld.

\World7.gdb\roads_rail feature class from ArcAtlas: Our Earth, 1996, courtesy of Data+ and ESRI.

\World7.gdb\us_cities feature class from ESRI Data & Maps 2006, courtesy of ArcWorld.

\World7.gdb\w_cities feature class from ESRI Data & Maps 2000, courtesy of ArcWorld.

\World7.gdb\world30 feature class from ESRI Data & Maps 2006.

\World7.gdb\WWF_Eco feature class from ESRI Data & Maps 2004, courtesy of World Wildlife Fund.

Data license agreement

Licensee shall not remove or obscure any copyright or trademark notices of ESRI or its Licensors.

Term and Termination: The license granted to Licensee by this Agreement shall commence upon the acceptance of this Agreement and shall continue until such time that Licensee elects in writing to discontinue use of the Data or Related Materials and terminates this Agreement. The Agreement shall automatically terminate without notice if Licensee fails to comply with any provision of this Agreement. Licensee shall then return to ESRI the Data and Related Materials. The parties hereby agree that all provisions that operate to protect the rights of ESRI and its Licensors shall remain in force should breach occur.

Disclaimer of Warranty: The Data and Related Materials contained herein are provided "as-is," without warranty of any kind, either express or implied, including, but not limited to, the implied warranties of merchantability, fitness for a particular purpose, or noninfringement. ESRI does not warrant that the Data and Related Materials will meet Licensee's needs or expectations, that the use of the Data and Related Materials will be uninterrupted, or that all nonconformities, defects, or errors can or will be corrected. ESRI is not inviting reliance on the Data or Related Materials for commercial planning or analysis purposes, and Licensee should always check actual data.

Data Disclaimer: The Data used herein has been derived from actual spatial or tabular information. In some cases, ESRI has manipulated and applied certain assumptions, analyses, and opinions to the Data solely for educational training purposes. Assumptions, analyses, opinions applied, and actual outcomes may vary. Again, ESRI is not inviting reliance on this Data, and the Licensee should always verify actual Data and exercise their own professional judgment when interpreting any outcomes.

Limitation of Liability: ESRI shall not be liable for direct, indirect, special, incidental, or consequential damages related to Licensee's use of the Data and Related Materials, even if ESRI is advised of the possibility of such damage.

No Implied Waivers: No failure or delay by ESRI or its Licensors in enforcing any right or remedy under this Agreement shall be construed as a waiver of any future or other exercise of such right or remedy by ESRI or its Licensors.

Order for Precedence: Any conflict between the terms of this Agreement and any FAR, DFAR, purchase order, or other terms shall be resolved in favor of the terms expressed in this Agreement, subject to the government's minimum rights unless agreed otherwise.

Export Regulation: Licensee acknowledges that this Agreement and the performance thereof are subject to compliance with any and all applicable United States laws, regulations, or orders relating to the export of data thereto. Licensee agrees to comply with all laws, regulations, and orders of the United States in regard to any export of such technical data.

Severability: If any provision(s) of this Agreement shall be held to be invalid, illegal, or unenforceable by a court or other tribunal of competent jurisdiction, the validity, legality, and enforceability of the remaining provisions shall not in any way be affected or impaired thereby.

Governing Law: This Agreement, entered into in the County of San Bernardino, shall be construed and enforced in accordance with and be governed by the laws of the United States of America and the State of California without reference to conflict of laws principles. The parties hereby consent to the personal jurisdiction of the courts of this county and waive their rights to change venue.

Entire Agreement: The parties agree that this Agreement constitutes the sole and entire agreement of the parties as to the matter set forth herein and supersedes any previous agreements, understandings, and arrangements between the parties relating hereto.

Installing the data and resources

Mapping Our World Using GIS includes a CD at the back of the book. Look for the one labeled Student Data and Resources. This CD contains the GIS data and other documents you will need to complete the projects.

Installation of the student data requires approximately 300 megabytes of disk space. Installation of the teacher resources requires approximately 25 megabytes of disk space.

Installing the student data

Follow the steps below to install the files. Do not copy the files directly from the CD to your hard drive. A direct file copy does not remove write-protection from the files, and this causes data editing steps in the projects not to work.

1. Put the CD in your computer's CD drive. A window like the one below will appear.

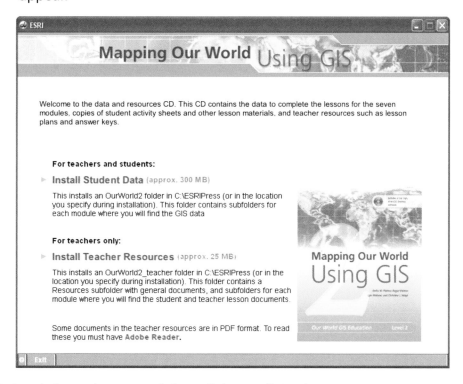

2. Read the welcome and then click Install Student Data.

3. Read the information on the next panel, then click the link. This launches the InstallShield Wizard.

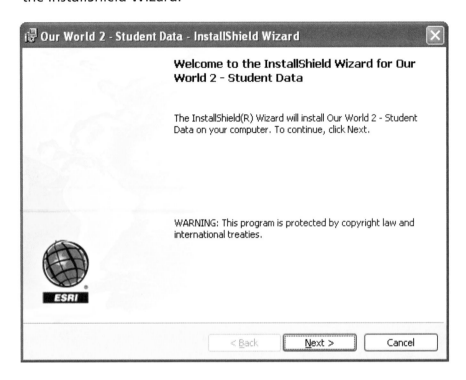

4. Click Next. Read and accept the license agreement terms, then click Next.

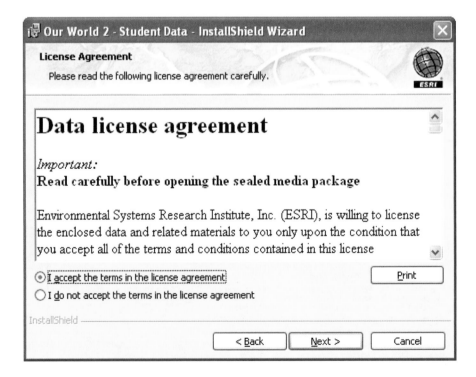

5. Accept the default installation folder or click Change and navigate to the drive or folder location where you want to install the data. Please make a note of where you install the data.

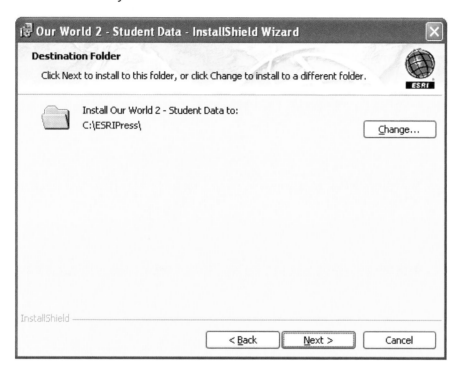

6. Click Next. The installation will take a few moments. When the installation is complete, you will see the following message:

7. Click Finish. The student data is installed on your computer in a folder called OurWorld2.

If you have a licensed copy of ArcGIS Desktop 9 (ArcView, ArcEditor, or ArcInfo license) installed on your computer, you (or your student) are ready to start *Mapping Our World Using ArcGIS*. Otherwise, follow the Installing the software instructions to install and register the trial software.

Installing the teacher resources

Follow the steps below to install the files:

1. Click Home at the top of the Welcome window if it is still open. Otherwise, put the CD in your computer's CD drive. A window like the one below will appear.

2. Read the welcome, and then click Install teacher resources.

3. Read the information on the next panel, and then click the link. This launches the InstallShield Wizard.

4. Click Next. Accept the default installation folder (C:\ESRIPress) or click Change and navigate to the drive or folder location where you want to install the teacher resources.

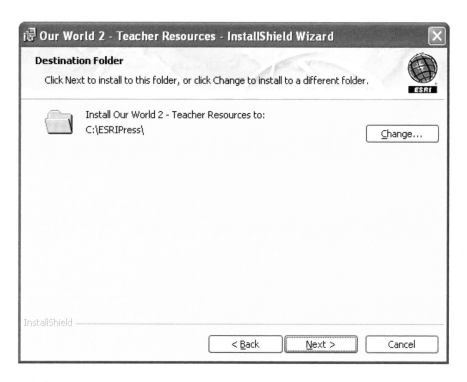

5. Click Next. The installation will take a few moments. When the installation is complete, you will see the following message:

6. Click Finish. The teacher resources are installed in a folder called OurWorld2_teacher.

Your installation is complete.

Uninstalling the data and resources

To uninstall the student or teacher and resources from your computer, open your operating system's control panel and double-click the Add/Remove Programs icon. In the Add/Remove Programs dialog box, select the following entry and follow the prompts to remove it:

OurWorld2 - Student Data and Resources

OurWorld2 - Teacher Resources

Installing and registering the trial software

The ArcGIS software included on this DVD is intended for educational purposes only. Once installed and registered, the software will run for 365 days. The software cannot be reinstalled nor can the time limit be extended. It is recommended that you uninstall this software when it expires.

System requirements

Before installing the ArcGIS Desktop 9.2 software, make sure your computer meets these system requirements:

- Microsoft Windows XP, Windows 2000, or Windows Vista operating system
- Disk space 1.2 GB
- RAM 1 GB minimum
- DVD drive (required for installation)

Note for users of Windows Vista

Service Pack 4 for ArcGIS 9.2 is required for Windows Vista support. There are some additional issues that you should be aware of before working with ArcGIS 9.2 on Windows Vista. Please visit this book's Web site (www.esri.com/ourworldgiseducation) for more information.

Install the software

Follow the steps below to install the software.

1. Put the software DVD in your computer's DVD drive. A splash screen will appear.

2. Click the ArcGIS ArcView installation option. On the Startup window, click Install ArcGIS Desktop. This will launch the Setup wizard.

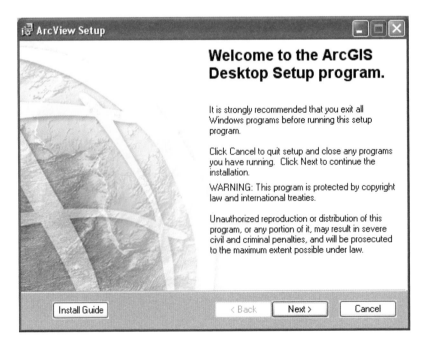

3. Read the welcome, then click Next.

4. Read the license agreement. Click "I accept the license agreement" and click Next.

5. The default installation type is Typical. You must choose the Complete install, which will add extension products that are used in the book. Click the button next to Complete.

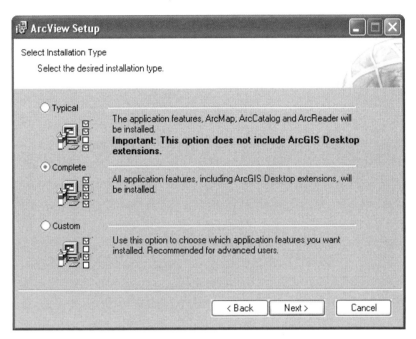

6. Click Next. Accept the default installation folder or click Browse and navigate to the drive or folder location where you want to install the software.

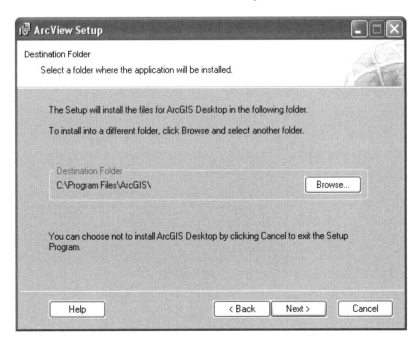

7. Click Next. Accept the default installation folder or navigate to the drive or folder where you want to install Python, a scripting language used by some ArcGIS geoprocessing functions. (You won't see this panel if you already have Python installed.) Click Next.

8. The installation paths for ArcGIS and Python are confirmed. Click Next. The software will take several minutes to install on your computer. When the installation is finished, you will see the following message:

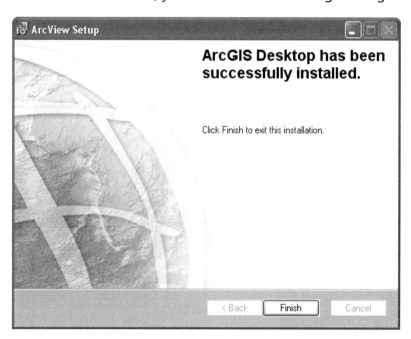

9. Click Finish.

Register the software

10. Locate your registration code on the software DVD jacket in the back of the book.

11. On the next panel, click Register Now and select the desired registration option. Be sure to register the three extensions used in this book (Network Analyst, Spatial Analyst, and 3D Analyst).

- *If you register automatically using the Internet, be sure to fill in the Organization field with a complete name; do not use an acronym or abbreviation or leave it blank. If you receive a message indicating a problem, contact ESRI Customer Service at 888-377-4575 to resolve it. Once the registration is complete you can start using ArcGIS Desktop.*

- *If you register manually, you will receive an authorization file in a *.esu9 format. Save this file to your computer. Open the Registration Wizard from the Desktop administrator and finish the registration process. Browse to the saved authorization file to complete the registration.*

Download and install the latest Service Pack

ESRI periodically releases software updates and corrections called service packs. The projects in this book require a minimum of ArcGIS 9.2, Service Pack 3 for all instructions to work properly. The following steps will guide you to download the latest service pack from the Internet and install it on your computer. *Note: You must have ArcGIS 9.2 installed and registered before you install the service pack.*

1. Connect your computer to the Internet and browse to http://support.esri.com.

2. Click the Downloads tab at the top of the page, and then click Patches and Service Packs.

3. Scroll down the list and click ArcView.

- ArcPad StreetMap (1 file)

- ArcReader (16 files)

- ArcSDE *(9.1 and older)* (55 files)

- ArcView (51 files)

- ArcView 3.x (42 files)

- ArcWeb Services APIs (28 files)

4. Click the link for the latest ArcGIS 9.2 service pack (Service Pack 4 was the latest one available at the time this book was published.)

> ▣ ArcGIS 9.2 Service Pack 4
> November 16, 2007 - SERVICE PACK
> This Service Pack contains performance improvements and maintenance fixes. Please download and install this required Service Pack at your earliest convenience.

5. Scroll down to the Installation Steps section. You will need to download the ArcView 9.2 Evaluation Edition files if you installed the trial software that comes with this book. (If you are using a fully licensed version of ArcGIS Desktop, download the ArcGIS Desktop—ArcView, ArcEditor, ArcInfo—files instead.) Click the msp link to save the files to your

> **ArcView 9.2 Evaluation Edition**
> ArcGISEvalEdition92sp4.msp   175 MB

computer.

6. Close your Internet browser. In your operating system's file browser, navigate to the saved msp file. Double-click the msp file to start the install process. When Setup starts, follow the instructions on your screen.

If you have questions or encounter problems during the installation process, or while using this book, please use the resources listed below. (ESRI Support Services does not answer questions regarding the ArcGIS 9 trial software DVD, the student data and resources, or the contents of the book itself.)

To resolve problems with the trial software or exercise data, or to report mistakes in the book, send and email to ESRI workbook support at learngis@esri.com.

To stay informed about exercise updates, frequently asked questions, and errata, visit the book's Web page at www.esri.com/ourworldgiseducation.

Uninstalling the software

To uninstall the software from your computer, open your operating system's control panel and double-click the Add/Remove Programs icon. In the Add/Remove Programs dialog box, select the following entry and follow the prompts to remove it:

ArcGIS Desktop

The ArcGIS Desktop software (ArcView license) accompanying this book is provided for educational purposes only. Once installed and registered, the software will run for 365 days. The software cannot be reinstalled nor can the time limit be extended. It is recommended that you uninstall this software when it expires.

The workbook and media are returnable only if the original seal on the software media packaging is not broken, the media packaging has not otherwise been opened, and the software has not been installed.

Purchaser may not redistribute the software or data packaged with this book. Libraries purchasing this book are subject to the following restrictions: Discs may not be loaned; software cannot be downloadable to anyone other than the purchaser; and discs must be removed before books are placed in circulation.

Please carefully read the instructions on the previous pages for important information about installing the data and software on the DVDs. You will also need to register the software and download and install files from the Internet.

To resolve problems with the trial software or exercise data, send an email to ESRI workbook support at learngis@esri.com.

For information about purchasing a fully licensed version of ArcGIS Desktop, visit www.esri.com/software/arcgis/about/how-to-buy.html.